Time Machine

Time Machine

Bernard Walton

BBC BOOKS

This book is dedicated to the memory of my parents, James and Else

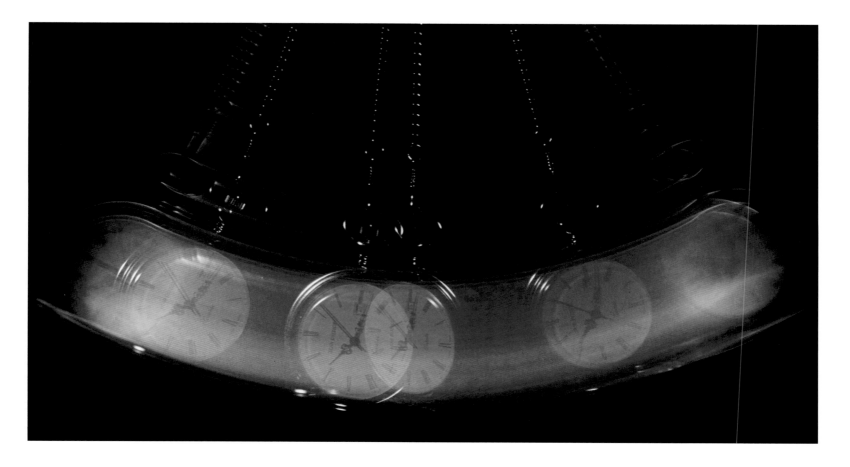

This book is published to accompany the television series entitled
Time Machine, which was first broadcast on BBC1 in 2004.
Executive producer: Sara Ford
Series producer: Bernard Walton

Published by BBC Books, BBC Worldwide Ltd, Woodlands,
80 Wood Lane, London W12 0TT

First published 2004
Text © Bernard Walton 2004
The moral right of the author has been asserted.

ISBN 0 563 48774 7

Commissioning Editors: Shirley Patton and Nicky Ross
Project Editor: Sarah Miles
Copy Editor: Caroline Taggart
Cover Art Director: Pene Parker
Book Art Director: Linda Blakemore
Designer: Martin Hendry
Picture Researcher: Frances Abraham
Production Controller: Arlene Alexander

Set in Gill Sans
Printed and bound in Great Britain by Butler & Tanner Ltd,
Frome and London
Colour separations by Radstock Reproductions Ltd,
Midsomer Norton

For more information about this and other BBC books, please visit
our website on www.bbcshop.com or telephone 08700 777 001.

Contents

Chapter 1
Time   12

Chapter 2
Earth   28

Chapter 3
Life   70

Chapter 4
Humans   114

Introduction

TIME IS ONE of nature's great mysteries. It is formless and intangible, yet we sense it around us everywhere. We see it passing as the second hand sweeps around the face of a clock, keeping a pace so constant that it can be divided into tiny increments and measured precisely. Time flows past us in one direction only. Our sense of the world and of being alive – our consciousness – happens in the present moment only, a moment so fleeting that it disappears into the past just as we experience it. It is as if time is a journey in which the present leaves the past behind and moves inexorably into the future.

We take time's relentless march forward for granted, but we forget that our perspective is distorted. Ours is a world of short and middling timescales – seconds, minutes, hours, days, years. The ticking of a clock and the setting of the sun are meaningful to us, as are the annual cycle of seasons and the growth of our children over years. But outside these familiar timescales, our intuitive grasp of time dissolves. Hummingbird wings are a blur to our minds because we cannot perceive time in hundredths of a second. Likewise, the Himalayas appear to be solid and still, though measurements tell us they are rising constantly.

The events and experiences that shape our lives are played out in hours and days. But the geological processes that shape the Earth can last millions of years. Compared to the vast expanse of geological time, a human life span is minuscule. Even in a thousand years little changes geologically – yes, a few volcanoes erupt, coastlines are eroded, and rivers may change their course – but these are minor changes. From a human perspective, the world seems solid and fixed. But nothing could be further from the truth.

If only we could change our perspective and watch how the Earth has changed over millions of years. If only we had enough time to see everything and to do everything, enough time to explore the past and look into the future. If only there was a way of playing around with time. What a thought! Yet this intriguing idea was proposed as long ago as 1895 in H. G. Wells's novel *The Time Machine*. His machine made the unthinkable possible – not only could it travel through our three-dimensional world, it could also travel in the fourth dimension: time.

Well now, imagine if there really was a time machine. A machine that could travel forward to the future or back to the past. A machine that could speed up time and also slow it down, compressing millennia into milliseconds and expanding milliseconds into minutes. Where would you go and what would you like to see? Without question, you would get an extraordinary insight into our whole world – past, present and future. You could watch changes in nature that are too slow for the unaided mind to comprehend. You could see continents sliding through the oceans, coalescing to form giants or splitting into islands. You could see whole mountain ranges rise from the ground and crumble back again, eaten away by nothing but weather. Or perhaps you would put the brakes on time and slow it almost to a halt, stretching out the seconds to watch the world frame by frame. Then you might learn how a hummingbird flaps its wings, or how an earthquake can unleash forces of such catastrophic intensity that they change the world in a split second.

By using a machine to crush or expand time, or to look back at a particular moment, you would be able to see and understand the hidden processes and forces of our world.

Our Time Machine

Time travel remains a dream, but we can visualize the experience by using images produced by a movie camera and computer graphics. The Time Machine of this book and the accompanying series consists of a camera and computer that create the illusion of real time travel.

If time as we see it is a distortion of reality, then certainly our Time Machine will give a distorted view too, and what we see when we use it will be limited by our imagination. But we can nevertheless place it squarely on the science we know to date. What we propose is a Time Machine based on what the camera can do with time; we will extend that ability into a 'deep time lapse'

On 18 May 1980, the eruption of Mount St Helens in Washington State, USA, was the largest ever recorded in North America. The volcano had lain dormant since 1857, concealing its highly dangerous disposition. Geological timescales are often far larger than we can comprehend, frequently to our cost.

In 1887, Eadweard Muybridge pioneered the use of high-speed photography and employed novel techniques to discover whether a galloping horse ever has all four feet off the ground at the same time. Previously, paintings showed horses galloping with their legs outstretched like a rocking horse, which was how most people then perceived them.

of centuries and millennia by using computer-generated images to create the experience of travelling in time.

So how can we simulate time travel? Well, despite appearances to the contrary, time is not truly continuous. Both space and time are quantized, which means they exist in tiny chunks (quanta) that cannot be divided. The smallest chunk of space is a Planck length, which is a hundredth of a billionth the size of a proton (or 0.00000000000000000000000000000001 centimetres long). The smallest chunk of time is a Planck time, and this is the time that light takes to travel one Planck length (to write it in seconds, you would need 43 noughts and a one after the decimal place). So in reality, time consists of a series of very short moments – much like the frames in a movie reel.

A camera captures moments of time that are usually no smaller than 1/20,000th of a second. A stills camera captures an image at the exact moment the button is pressed. The shutter opens and closes and the film records the image. The instant the picture is taken, it becomes an image of the past, and it grows older as we look at it, because we are constantly moving forward in time. It is a view of the world in another time.

A movie camera can do more with time than a stills camera. Twenty-five (24 in the USA) still images, or frames, are captured every second the camera is turned on. Each frame is a moment of time, and as long as the film is played back at the same speed, it gives us, the viewers, the perception of a fluid movement in time and space.

The altering of time occurs when the film is speeded up or slowed down during recording or playback. Capturing 50 frames every second and playing them back at the standard speed of 25 frames a second means that everything is slowed down by half. On the other hand, if a camera captures 12 frames per second and the film is played back at 25 frames per second, the action is speeded up to twice the normal speed. Filming everything really slowly like this is called time-lapse photography. Playing around with the speed of the film as it goes through the camera means that when we view the film later we will see time pass at different speeds. It also means that we can not only speed up and slow down time, we can also reverse it

All the Earth's history in just one day

Our Time Machine can do more than just play with the speed of time, it can also compress time so that epochs in the Earth's history become a comprehensible period. If we squeeze a million years into seconds, we will see major events happen over a very short time. To help us to achieve that we can compact the lifetime of the planet into one day and represent this in a 24-hour clock. If the Earth started to form 4.55 billion years ago, we will treat that as midnight (00:00) at the start of a new day, and the present as the moment when the clock is striking midnight 24 hours later (24:00). In our Time Machine, therefore, one million years are compressed into just 19 seconds. Most of the living creatures and landscapes of today appeared in the last couple of hours before midnight, with humans arriving at just one minute to midnight.

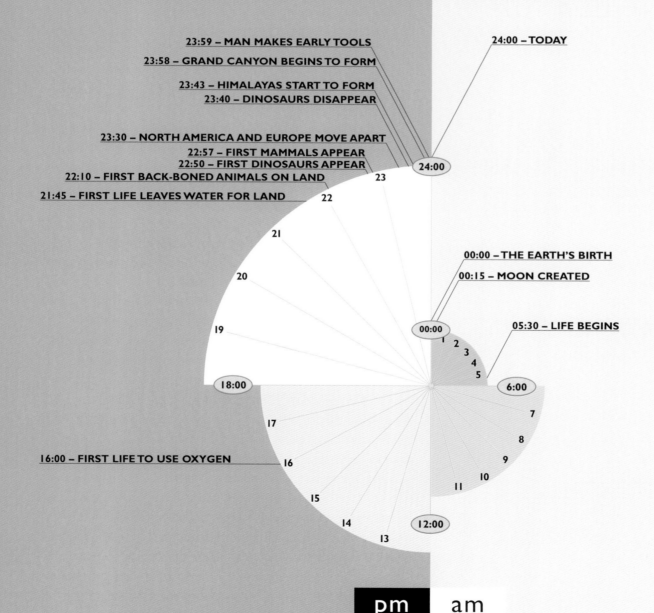

23:59 – MAN MAKES EARLY TOOLS
23:58 – GRAND CANYON BEGINS TO FORM
23:43 – HIMALAYAS START TO FORM
23:40 – DINOSAURS DISAPPEAR
23:30 – NORTH AMERICA AND EUROPE MOVE APART
22:57 – FIRST MAMMALS APPEAR
22:50 – FIRST DINOSAURS APPEAR
22:10 – FIRST BACK-BONED ANIMALS ON LAND
21:45 – FIRST LIFE LEAVES WATER FOR LAND

24:00 – TODAY

00:00 – THE EARTH'S BIRTH
00:15 – MOON CREATED
05:30 – LIFE BEGINS

16:00 – FIRST LIFE TO USE OXYGEN

pm am

at different speeds and even make it stop. In effect, the movie camera becomes a visual time machine. If the camera gives us the illusion of reality then, in conjunction with a computer, it can be used to give the illusion of time travel.

The camera can capture the movement of the slowest of things – a crawling snail, the opening of a flower at dawn – and make them all move very quickly when played back. Seeing the tendril of a bindweed spin in mid-air, looking for anything it can cling to so that the plant can continue to grow to greater heights, gives us an insight into how very slow objects move and how living things grow and develop.

We can also put time on hold, freeze it for a moment and then move around an object that is frozen in time. This technique is called 'time slice' and it requires a bank of several still cameras, sometimes as many as 90, lined up so that they all take a picture at the same time. If we then view the results as a series of images on a film, we get the sense that we have frozen time, but we can still travel through space to see the object at different angles. Time slice was made famous in the movie *The Matrix*, when the hero escapes bullets by dropping himself backwards. It gives a three-dimensional view of the frozen action and produces a multiple perspective of what is happening at that precise moment in time.

The film camera is a perfect time machine for showing many things, but it is limited to the period of time that it is set up to cover. Time-lapse techniques rarely cover more than a year. But with the help of a computer, we can extend that period to millions of years. New software enables us to show how real landscapes changed over greater periods of time. By laying out snapshots of time over a vast period, we can see the changes more easily.

Travelling in time

With the illusion of time travel complete, using computer graphics and camera techniques, our Time Machine can set off on its journey. In Chapter One we explore the curious nature of time and look at humankind's attempts to understand and measure it, from the first hunter-gatherers, who lived by the daily rhythm of the sun, to the modern era, where atomic clocks allow us to carve up and track every second with astonishing precision.

In Chapter Two we travel back in time to find out how the Earth has changed over its 4.55-billion-year life span. Compared to the other planets in the Solar System, the Earth is very peculiar. There is a mysterious absence of craters, and the Earth's surface is a jumble of mountains, highlands, canyons, and vast oceans. How did our planet become so unusual? The answer lies in a combination of two factors: the slow churning of molten rock deep within the planet, which continually stirs and buckles the Earth's crust; and the amazingly corrosive power of weather and water, both of which are driven by the power of the sun. Some of the changes these forces engender take less than a minute, others take millions of years.

Chapter Three explores another unique feature of our planet: life. By adjusting timescales in the Time Machine from milliseconds to millennia, we can explore the amazing changes that life makes in its fight for survival and the huge impact these have on the way our planet looks. We will see how the tiniest early forms of life – cyano-bacteria – transformed the atmosphere and turned the sky from orange to blue, recolouring the entire planet. We will speed up evolution to see why the female hummingbird was forced to change the shape of its bill, and we will zoom in on one small, nervous forest animal to see how time tranformed it into the modern horse.

In the final chapter the Time Machine will show us how our own species has changed the world. In a geological instant we tamed the wild and built huge cities. Civilizations rose and fell, and natural habitats were cleared to make way for new landscapes of concrete and steel. With each step forward our progress accelerated, bringing agriculture, the industrial revolution, and finally the era of high technology and communication that has transformed the world into a global village.

For humans, time is now at a premium, as we are slowly destroying the Earth. Natural resources are running out and we are being forced to seek new solutions to our energy requirements. In the frenzied drive for greater speed and efficiency, we constantly strive for faster means of production, consuming ever more. Under such conditions, how much time does our global civilization have left?

Although most people perceive the cyclic rhythm of time, it has different meanings for different individuals. For instance, among the Australian Aboriginals there is no such thing as just the past – they combine past and present into one. The Hopi people of North America do not believe it is right to look into the future. For many cultures around the world, timekeeping is not important, mainly because planning is not a part of their lives. Things happen when they happen, irrespective of time.

So what exactly is time? St Augustine of Hippo, a great fifth-century philosopher, gives an observant answer: 'If no one asks me, I know; but if any person should require me to tell him, I cannot.' Even today many scientists are baffled by its complex nature. Just when we think we have grasped it, it slips out of our hands. Most dictionaries define time as the period or interval between two separate events in space; where the 'period' is defined as an interval of time between two separate events – a circular definition, but probably the best there is. And does time really exist? We cannot see, smell, feel, taste or hear it. Yet we can perceive its effects passing through our senses, as we glimpse a golden sunrise or smell a spring flower – sensations that last only a moment and then are gone.

LEFT: We all sense time passing as we live through each day, yet not everyone agrees as to what exactly time is.

RIGHT: The cheetah is the fastest creature on land. To be a successful predator, it has to have an acute sense of timing.

Time

Measuring time

It is not just our modern world that has become obsessed with time. Measuring time was also important in the earliest human societies for religious events, political meetings and city planning. So how did this obsession begin?

The way we perceive the transition of time varies enormously according to what we are doing. As Einstein once said, an hour spent in the dentist's chair seems longer than the same time spent with a pretty woman. When we are busy, time goes by quickly, and when we are bored or looking forward to a happy event, like a holiday or a wedding, it goes very slowly. This, of course, is totally subjective. One person's sense of time passing is often very different to another's. In reality, most of us have a poor idea of the passage of time.

We can get a better grip on the passage of time by measuring it. From the beginning of human history, time has been measured by the movement of the Earth

A water clock from Karnak Temple, Egypt, dated 1415 BC. Clocks such as this one were not only decorative ornaments, but also practical tools for measuring time. The clock would have been filled with water, which was then left to trickle out through holes in its base, the flow marking the passing of time.

relative to the sun and stars. Early thinkers such as Aristotle, who studied the heavenly bodies, perceived time as motion. Most timekeeping used some form of moving device, something that changed regularly and predictably, like the sun and, later, hourglasses.

The earliest known timepiece is the shadow clock, estimated to date from about 3500 BC and known as a horologium to the Greeks, who ascribed its invention to the Babylonians. The most common form is a gnomon, a vertical pole or obelisk that casts a moving shadow as the sun moves across the sky. An Egyptian shadow clock of the eighth century BC still exists. In the third century BC the Chaldean astronomer Berossus described the first hemispherical sundial, which worked on the same principle as the gnomon.

The 24-hour day

The concept of having 24 hours in a day was devised by Egyptian and Greek scientists. The whole day was split into day *time* and night *time*, and the day was then divided into 12 parts. No one is sure why, but one theory is that these parts may have represented the 12 signs of the zodiac which measured the calendar months. Alternatively they may have originated in the Egyptian shadow clock or sundial, which divided a sunlit day into ten parts plus two 'twilight hours' in the morning and evening.

However, not every place could rely on continuous sunshine to tell the time, so other ideas had to be thought up. A notched burning candle telling the hours was quite a reliable way of keeping track of time, as was the Chinese practice of burning a knotted rope. By noting the length of time necessary for the fire to travel from one knot to the next they could keep a note of time. The hourglass is an ancient timepiece, too. It is a simple device based on the time it takes sand to flow from one glass chamber to another through a thin hole.

Another popular instrument is the water clock, or clepsydra, in which the flow of water measures the passage of time. Its original and simple form was a vessel with several little openings at the bottom, through which the water escaped. It was used for measuring the time during which people were allowed to speak in the courts of justice at Athens. As early as 270 BC, the inventor Ctesibius of Alexandria started using more sophisticated devices involving gears, so that the time could be told by reading a dial. In due course, the flow of water was replaced by a weight falling, paving the way for the invention of the mechanical clock. (The name *clock*, which originally meant *bell*, came from the bells that were rung out to tell the time of day in the late Middle Ages, usually calling people to prayer.)

The mechanical clocks of the fourteenth century were big and cumbersome. Henry de Vick of Württemberg built one for the royal palace in Paris and it was powered by a 227-kilogram (500-pound) weight that slowly dropped a distance of 9.8 metres (32 feet). The mechanics were crude and the clock was wildly inaccurate. Clocks of that era had dials with only one hand, which told the time to the nearest quarter-hour.

By the sixteenth century, however, the constant swing of a pendulum had been noted by Galileo. By the middle of the seventeenth, Dutch physicist Christian Huygens had worked out how a pendulum could be used both to power and to regulate a clock. Ten years later English physicist Robert Hooke invented an escapement, which meant that the pendulum could be reduced in size. So, at last, clocks became smaller.

In the meantime, an ingenious way of powering timepieces had been invented in Italy, where, in about 1450, coiled springs were introduced as a replacement for the pendulum. Around 1500, Peter Henlein, a locksmith working in Nuremberg, Germany, created the first portable timepieces, known popularly as Nuremberg eggs. By 1660, Robert Hooke increased accuracy by introducing a spiral hairspring and in 1765 British inventor Thomas Mudge invented the balance wheel. These important elements of the watch are still used today.

Accuracy

As time measurements became more accurate and more important in our lives, a major change was made to the familiar clock face. The minute hand was added, allowing us to measure time intervals smaller than a quarter of an hour. The 12 hours that already appeared on the clock

The oldest surviving mechanical clock in the world is found in Salisbury Cathedral, England. Made in 1386 and originally housed in the Bell Tower, it was created to strike the hours, probably calling people to prayer. The cogs and wheels seen here are still used in mechanical clocks and watches today.

face served to mark five-minute intervals, so that one revolution of the minute hand took 60 minutes and, in turn, moved the hour hand on one hour. Later, a third hand was added that revolved once every minute, making 60 seconds within a minute.

In the eighteenth century accuracy was further improved by using jewelled bearings to reduce friction and prolong the life of both clocks and watches. Every town had to have a public clock by which everyone set their time. By the mid-nineteenth century, watches were small enough to be put on the wrist. Our lives were changing forever, becoming more ordered and scheduled. Knowing the time was an important part of everyday life: factories needed to tell their employees when to start and finish work; trains and ships required everyone to know when they were leaving and arriving.

The twentieth century brought electrical watches and, in 1929, the first quartz clock produced more accurate timekeeping than had ever been possible before. By the late 1950s, electronic watches with LEDs (light emitting diodes) and LCDs (liquid crystal display) appeared on the market, giving us digital reading of time, not only to the second but also to 1/100th of a second.

All of these sophisticated instruments serve us very well, but the problem is that no two timepieces can give exactly the same reading because no two devices are identical. Even today it is impossible to make identical mechanical clocks. Quartz watches measure time quite accurately, but over time they drift out of synchronization, losing a second every week. Synchronization became crucial when the increasing popularity of rail travel made it necessary to know the exact time at two different locations at the beginning and end of a journey. Until the nineteenth century, there was no standard time across England – the time in Penzance in Cornwall, for instance, was set locally by the midday sun and, because Penzance is further west than London, when it was 12 noon in London, it was 11.30 a.m. in Penzance. For the benefit of the traveller, the clock at Paddington Station in London had to be synchronized to that in Penzance. One master clock had several slave clocks that kept identical time using telegraphy. Thanks to the railways, all clocks around the country soon became synchronized. It took early radio broadcasters, such as the BBC, to tell people the exact time, allowing everyone to set their clocks and watches by listening to the 'pips' on the hour.

To avoid confusion across the world an international agreement was made to use noon at Greenwich, along the longitudinal line 0 degrees, as the starting point for calculations of time zones. Each time zone across the world has its own (usually hourly) offset from Greenwich Mean Time (GMT). In 1925 the numbering system for GMT was changed so that the day began at midnight. In 1928 the International Astronomical Union changed the designation of the standard time of the zero meridian to Universal Time, a term more widely used among scientists and astronomers. Originally GMT was set by the Earth's rotation, though as we shall see this was later found to have problems of its own.

Reliable timekeeping

As technology developed and split-second timing became essential for more elaborate machines, the international community wanted a more accurate way of measuring and keeping time around the globe. They found it in the atomic clock. The big difference between a standard clock in your home and an atomic clock is accuracy. The oscillation in an atomic clock is between the nucleus of an atom and the surrounding electrons. This oscillation is quite like the balance wheel and hairspring of a clock-work watch, but these sub-atomic oscillations are the most accurate beats of time and so the best natural timepieces known to us.

Up to the mid-1950s, the scientific standard of time, the second, was based on the Earth's rotation and was defined as 1/86,400th of the mean solar day. But then it was realized that the Earth's speed of rotation was quite irregular, sometimes slowing down and at other times speeding up, simply because the Earth wobbles naturally in space like a child's top. The Earth, it was discovered, was an inaccurate timepiece and so it became necessary to find a more reliable definition for the second.

So in 1955, the National Physical Laboratory in England built the first caesium-beam clock, to be used as a calibration source to which all clocks in the country were to be set. By international agreement, the second is now defined as 9,192,631,770Hz – that is the same number of beats per second as a resonating caesium atom. Over the next decade, more advanced forms of these clocks were created, and in 1967 the 13th General Conference on Weights and Measures defined the international second (SI) on the basis of vibrations of the caesium atom. The world's timekeeping system no longer relied on the inaccurate rotation of Earth around the sun. The long-term accuracy achieved by modern caesium atomic clocks is better than one second per one million years. Atomic clocks have increased the accuracy of time measurement about one million times in comparison to astronomical techniques.

Who is keeping the correct time?

There is no single master clock for the whole planet. Instead, the international time standard is based on the average of some 260 atomic clocks around the world. In the UK it is the National Physical Laboratory (NPL) that maintains and develops the national time standard. Astonishingly the NPL keeps the UK's time accurate to within one second in three million years, making any error almost infinitely minuscule.

The international time standard held by NPL and the other time laboratories is made freely available by their radio signal, which is picked up by specially designed radio clocks. The Rugby Radio Station, based in Rugby in Warwickshire, broadcasts a time signal all over the UK using NPL's atomic clocks. Clocks that pick up this signal can be bought on the high street, allowing anyone to share in this split-second timekeeping. Today when you call the speaking clock or hear the 'pips' on the radio, the time comes from the atomic clocks at Rugby or some similar device elsewhere in the world. The implications for our lives are enormous: video recorders remain accurate and will even switch automatically to daylight saving in autumn and spring every year. All computers linked to the Internet are also synchronized to the same atomic clock, allowing us to communicate with lightning speed and accuracy.

Today, accuracy is crucial for communication in a world where computers and telephone systems rely on

The first atomic clock, built in 1955, was the size of a dining room table. It was invented after the discovery that the rotation of the Earth was an inaccurate method of measuring time. Atomic clocks use pulses or oscillations of caesium atoms to measure time reliably – they have revolutionized modern timekeeping.

total time synchrony. Without atomic clocks, GPS (Global Positioning Systems) navigation, for pinpointing our position on Earth, would be impossible, computer networks would not synchronize and the position of the planets would not be known exactly enough for space probes and landers to be launched and monitored. All depend on synchrony that is driven by pulses of time.

Which way does time travel?

At first glance, that seems like an easy question – it only goes forward. Just as rivers flow downhill to the sea, not the other way around, so we are all getting older and we are ageing in the process. Time is driven forward by the fact that everything in the universe is moving from a highly ordered state to a less ordered state and eventually into total disorder. Rocks crumble or dissolve. Cars rust, paint discolours and flakes, buildings eventually fall down. This inevitable progression towards chaos is called the arrow of time.

So time has an effect on everything, and everything is going in one direction only – the direction of disorder and death. This may seem depressing, but our planet is in the unique position of being able to move some things from disorder to an ordered and structured state. We can make new things by reversing the process of decay. When we manufacture steel, for example, we are producing something that has a higher ordered state than its raw materials. The same can be said of glass, clocks, cars and computers. But this renewal can be done only at a local level and only by applying energy to the process. Energy is always required to reverse the procedure of disorder. Ironically that energy is usually the product of disorder, like heat from burning gas.

In fact, producing a higher ordered state is what living organisms do all the time – they generate complex chemical structures that allow them to exist (but this reversal of disorder does not reverse time). Proteins, hormones, electrical pathways, blood and bone structures are all highly ordered states of simple atoms, largely carbon, hydrogen and oxygen. Living things, however, must be able to maintain and repair themselves from decay and damage by renewing themselves. Nothing on the planet can do this except life. But sustaining life on the planet is possible only thanks to the massive amount of energy that hits it from outer space, in the form of light from the sun.

When it reaches the Earth, light is converted into new forms of energy. For green plants, this means using photosynthesis to produce biochemical energy, which is then passed through the food chain from the lowly plants to the top predators. But this reversal of disorder will happen only as long as the sun is there. Eventually and sadly the sun will burn out and so will life on our planet. However, that date is thought to be in about 4 billion years' time, so we still have plenty of time to enjoy ourselves.

When the sun's light hits non-living things, such as rocks, most of its energy turns into heat. However, heat dissipates in one direction only – from hot to cold. On a universal scale, this means that everything is moving from the hot Big Bang into the colder outreaches of space. This process will continue over vast amounts of time until the universe, and everything within it, becomes evenly spread out in space and temperature will be absolutely uniform. This will be the time when time itself ends and so does the universe. So where are we now in the timescale of the universe's existence? The origins of time's arrow can be traced to the Big Bang, which happened some 14 billion years ago. It is estimated that we are currently halfway through the universe's life. So, in about another 14 billion years the universe and everything in it will be truly on the way to its final heat death. Our solar system was created some 10 billion years after the Big Bang and it will disappear well before the end of the whole universe.

All this tells us that in our universe time is going only one way. But if we are to ponder the possibilities of time travel, we have to understand how time and space are interlinked.

The arrow of time is constant. Time goes in but one direction, which means that we only see rivers and streams flowing downhill or forwards, and never in reverse. Quite simply, in the same way as we never see nature's forces going backwards, so we never perceive time moving backwards.

The universe and the laws of nature set time. Each day starts with a sunrise in the east and ends with a sunset in the west. The spin of the Earth means that the sun can be accurately predicted to be in a certain place in the sky at a particular time, making it one of our oldest timepieces.

Can we really travel through time?

A hundred years ago, Albert Einstein came up with his famous general theory of relativity, which changed our perspective of time and space. Einstein calculated that time slows down when we travel. So time begins to 'warp' when one observer is travelling – in a rocket or aeroplane, perhaps – relative to another who is stationary down on Earth. This warping of time is called the 'time dilation effect' and it was tested in 1971 when two American physicists flew precision atomic clocks in aeroplanes round the world. They discovered that the time in the aircraft had slowed down relative to the ground-based atomic clocks by 59 nanoseconds (59/1,000,000,000th of a second) – exactly the time difference predicted by the mathematics used in Einstein's theory. This means that astronauts are relatively younger than most of us because time passes more slowly when they are travelling.

The greater the speed at which we travel, the greater the change in time. To witness the alterations of time you would have to travel close to the speed of light, which is impossible at present because of the amount of energy required. But if it were possible and a traveller journeyed to an outer star and back again within two years, when he returned he would find that 14 years had passed by here on Earth. In other words, if the traveller had a twin, the Earth-bound sibling would now be 12 years older than the traveller. This means, in theory at least, that if we were travelling very close to the speed of light, we could fast-forward in time, reaching the year 3000 in one year.

And what about going back in time? As anyone who has seen the film *Back to the Future* will know, this has its problems. Since everything that is happening now is determined by the past, going back and altering something could change the present. If you were to go back in time and save Marilyn Monroe from her drug overdose, she might be still alive today. This would either change the present or perhaps create a second Marilyn living in a parallel but separate universe or existence from our time. The logic of time travel can lead to extraordinary conclusions.

Stephen Hawking, author of *A Brief History of Time*, proposes a 'chronology protection hypothesis' that prevents time travel to the past. This is based on the assumption that we could never physically make a time machine because it would self-destruct as soon as it started to travel back in time. In fact, if we could go back in time to our younger years we would see duplicates of

ourselves. Hawking also maintains that we will never travel back in time for the simple reason that we do not see our descendants coming to visit us. But not everyone agrees.

What is 'Now'?

Current thinking by Julian Barbour, an English physicist, is that all time – present, past and future – exists all the

Albert Einstein was the first to make the connection between space and time. His theory showed that time was not a fixed constant, as Isaac Newton saw it, but was something that changed depending on how fast you travelled through space relative to a fixed position.

time. He says, 'My basic idea is that time as such does not exist. There is no invisible river of time. But there are things that you could call instants of time, or "Nows". As we live, we seem to move through a succession of Nows, and the question is, what are they? They are arrangements of everything in the universe relative to each other in any moment – for example, now.'

Barbour believes that the illusion of time passing is perceived only by living things and as such we perceive it as if we were moving on a journey through the present. If time is in fact moving into the future, then any choices we are about to make have already been determined in a matrix of different existing Nows. This seems to suggest a somewhat fatalistic progression to our existence, but Barbour maintains that we have real choice, because there is a multitude of different possible outcomes to every eventuality. This would mean we could travel through different Nows and experience what we would call past events, without altering the present or the future, because there are so many different Nows. The consequence of any action or choice made by us would bring us to new Nows.

So, according to current thinking, there can be no meaningful concept of time as we see it, with a now, a past and a future. Most scientists dabbling in time think that time as we perceive it is an illusion. It certainly exists in some form; but perhaps not as we know it or experience it. Past and future are there; it is a matter of knowing not so much where they are, but perhaps what they are. So if all of time exists in some way, whether or not we see it as past, present and future, travelling through time should be a possibility. Certainly Einstein's general theory of relativity does not rule it out and it has led to some interesting new observations and theories about time.

The possibilities of time travel

It was only in the 1980s that a concept of travelling through time resurfaced through the idea of wormholes, known also as stargates. These are time tunnels through space, resembling black holes in that they are regions of intense gravity, but instead of being a portal to nowhere, wormholes have an entrance and an exit. They are short

Time machines and time travel

Although Einstein put time firmly as the fourth dimension, the idea had been around a long time. In fact it came out of a student debate at Imperial College London attended by H. G. Wells. Subsequently, in 1895, at the age of 28, he wrote his famous novel about the exploits of a Victorian scientist lured by the idea of time travel. It was the golden age of science. Wells's time machine was a precision-made carriage that physically took his hero, the Time Traveller, to the future and back again.

Wells was more intrigued by the future of mankind than by the time machine itself, and his vision was inspired by the sharp class divisions in Victorian society. His hero travelled to the year 802,701 to discover a nightmare world where evolution has tranformed the classes into separate species. He found himself among a frail and graceful race called the Eloi, who lived in a faded garden of Eden where all their needs were met and there was no need to work. The Eloi seemed a pale shadow of their middle-class forebears – they had become permanently idle and had lost the ability to reason or concentrate, as though their idyllic home had made intellect redundant. The Traveller soon discovered why. The Eloi were merely fatted calves, tended and then eaten by the Morlocks, a savage, apelike branch of the human race who lived underground. Horrified, the Traveller returned to his machine to escape, fighting off Morlocks on the way. He set off for the future again, stopping 30 million years later to witness a deathly landscape in which the sun barely shined. Then he returned to the present and to the dinner-party guests who had come to see his invention.

As well as launching Wells's career as an author and inspiring the twentieth-century genre of science fiction, *The Time Machine* was a landmark in considering time as a medium through which one could travel. Wells's scenario bore an uncanny resemblance to what was to come in our understanding of time. After all, the book was published less than ten years before Einstein's general theory of relativity, which was to lead to a real possibility of time travel.

H. G. Wells's novel, *The Time Machine*, was the first fiction to explore the possibility of real time travel. The book became a film in the 1960s, starring Rod Taylor. Ever since there have been many other imaginary adventures in time travel, both in novels and in films.

cuts between two distant positions in space. Anything entering a wormhole will go to different places and times, past or future, depending on the direction in which it is travelling. This wormhole route through time and space was made famous by Carl Sagan's book *Contact*, which was later made into a film of the same name.

In theory, it should be possible to build a time machine that can take you to the past, but the technology and power required by such a machine are likely to remain permanently beyond human capabilities. US physicist Frank Tipler proposed a time machine in the 1970s. The Tipler Cylinder is a superdense spinning cylinder that warps the fabric of space–time. The only problem is that the cylinder would have to be infinitely long and would weigh more than the sun.

Another suggestion is a time machine made out of cosmic strings – vast, theoretical structures left over from the early universe. But this would be no easier than building a Tipler Cylinder. Even if a future civilization managed to build such a machine, there remains the objection that future time travellers would already have come back to visit us. But according to Stephen Hawking, no time machine could travel further back than the date of its creation.

In recent years, another means of travelling in time has emerged. Advanced civilizations of the future could use vast supercomputers to model the world in much the same way as modern weather computers model the atmosphere. Assuming computing power continues to grow at its current rate, the future computers could recreate Earth's history in all its detail – including the lives of everyone who has ever lived. This theory, although perhaps taking too mathematical an approach, is considered seriously by physicists, but its implications are disturbing. If computers of the future can generate replicas of the past with such ease, who is to say we are not living in one of the replicas?

Clocks are essential in our busy modern lives. Universal time has been set so that we can coordinate our plans as well as our machines, no matter where we are in the world. Time has become so crucial for structuring our daily lives that many feel we are slaves to time and becoming increasingly so as our technology advances.

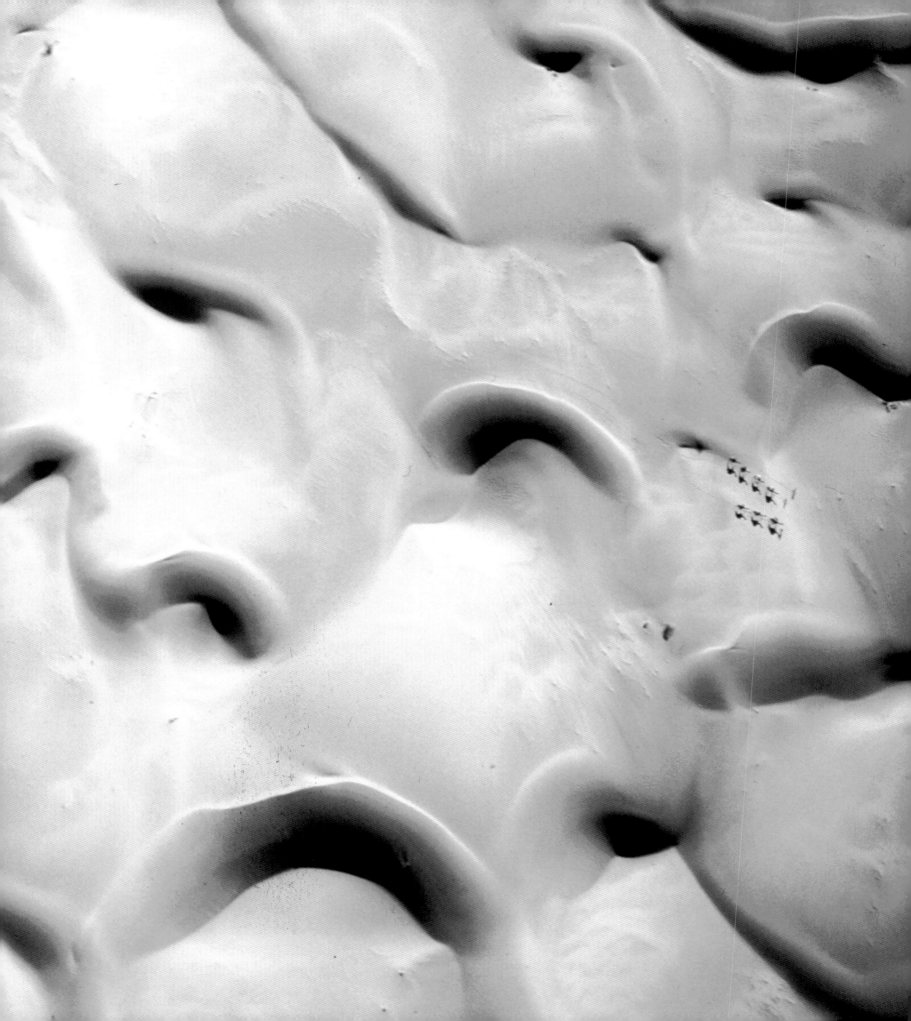

Look around and see the landscape that surrounds you. Everything about it looks solid and timeless. Even a place you have known for most of your life never seems to change. Occasionally a river bursts its banks and changes its course slightly, with a sand bank appearing and then disappearing, but generally a photograph of a mountain or a stream taken a hundred years ago will look much the same as it does today. Most natural landscapes rarely change at all – except, of course, when we alter them for our own use. They seem to have remained intact since the beginning of time.

But these permanent-looking features have misled us about the true dynamism of our planet. Changes in the geological landscape span many thousands or even millions of years and so, to us, they are almost imperceptible. Our life span is very short in comparison to the life of a river or mountain – so short that we see only a glimpse of the geological process. To gain some perspective, we need to crush time so that those changes happen in milliseconds rather than millennia. With the help of the Time Machine we can get a different view, not only of the forces that bring about change, but of the reasons for their existence in the first place.

At first glance the world looks solid, as if it had been there forever. Yet there are clues to its changing faces. We see sand dunes shifting in the deserts (left) and glaciers moving within a few hours (right). If we travelled quickly through time beyond our own lifetimes, we would even see solid mountains and continents move.

2 Earth

2:1 Restless planet

How old is the Earth?

What sort of timescale are we talking about when it comes to the age of a mountain or a river? How old is the Earth? How did we find out?

The age of the Earth and the length of time we humans have been on it have always been curiosities to us. For a long time there was very little to go on, but it didn't stop people trying. In 1650, Archbishop James Ussher in Ireland calculated that the Earth was created in 4004 BC, which would make it about 6000 years old. He made this calculation using the Bible as his reference, going back systematically through generation after generation of the people whose ancestry was documented there. He was able to work right back to Adam, the first man created by God. For those who believed in the very word of the Bible, this gave a clear picture of the length of time we had been on Earth. But the Christian Church was soon to be in for a huge shock when this calculation was disputed by scientific and rational deduction.

Reading the rocks

During the seventeenth and eighteenth centuries, science underwent a long and successful renaissance, giving fresh meaning to the world. A new vision of the Earth was beginning to link things that had previously been seen as separate, like force and time. Sir Isaac Newton's Laws of Motion, published in 1686–7, explained the concept of gravity and totally revised the way scientists thought about moving bodies. The Industrial Revolution was on its way, and mining was at its peak: scientists were taking an interest in rocks both for their usefulness and for a better understanding of the Earth itself. It was believed that the age of the Earth could be estimated by studying how rocks were formed. Geologists started to read rocks very much like a history book, starting with the present, which was usually seen at the surface, and then digging deeper and deeper into the past. However, it was not that simple: rock was rarely laid down in neat layers reflecting the passing of time as you descended through them. It started to look more like a puzzle book than a history book.

These mountains in Teton National Park, Wyoming, USA, appear to us as a timeless landscape. If we could compress time a million times, to see the true dynamic forces of the Earth at work, we would then witness such landscapes being created, and eventually destroyed.

It was James Hutton, a Scottish farmer and businessman with a great interest in the natural world, who threw a new light on how old the Earth was. In 1788 he made a discovery that was to change our view of the planet. Fascinated by the rock formations at Siccar Point, near Edinburgh on the east coast of Scotland, he calculated that at least seven geological events had happened in this one place. There were dramatic differences between these events, from the laying down of sedimentary rock to lava flows. Often the newly formed rocks had been twisted from a horizontal position to a vertical one by upheavals in the Earth. Hutton pondered that this must have taken huge, even limitless, periods of time to complete. He proposed that the Earth had been here always and infinitely. However, many scientists were not happy with this explanation and thought that the Earth had a real beginning. The problem was to find a way of calculating it using the current state of science.

The theories move on

In the middle of the eighteenth century, William Thomson, who later became Lord Kelvin, was researching thermodynamics, the behaviour of heat, at Glasgow University. He heard reports that as coal miners dug deeper into the Earth, they discovered that the rocks were much hotter than those on the surface, indicating that there was a heat source deep inside the Earth. Kelvin considered that this heat was left over from the creation of the planet and was still dissipating out from the Earth's core. If this was the case, then there was a very good chance, he thought, that he could use the laws of thermodynamics to calculate the age of the Earth. He knew that he was making a lot of assumptions in his calculations. He supposed that the only source of heat within the Earth was created when the planet was

formed. He suggested that the Earth was cooling by conduction, in the same way that a central-heating radiator cools down after it is turned off. He assumed that the temperature of the Earth's atmosphere had remained the same throughout its history. He also suggested that the temperature inside the Earth was the same as that of a volcano – 1100°C (2000°F). He then worked out the heat-retaining properties of the Earth and obtained the latest estimation of the planet's size. Finally, using the temperature gradient measured in the coal mine, he calculated how long a body the size of Earth would take to cool.

As a result of all this, Kelvin worked out that the Earth was somewhere between 20 and 400 million years old. Over the next 30 years, with the help of other scientists, this was refined to something closer to 20 million years. But again, not everyone was happy with the conclusion.

Towards the end of the nineteenth century, an Irish geologist, John Joly, at Trinity College, Dublin, tried another method of ageing the planet. He began by calculating how long it took the sea to become salty, assuming that it was fresh water to start with. By estimating the amount of salt deposited by all the rivers in the world into all the seas he calculated that the age of the planet was about 99 million years. By the beginning of the twentieth century most geologists were reaching a figure of about a hundred million years. However, science still needed something more conclusive, something simpler, based on as few assumptions as possible.

It was only in 1902, when Ernest (later Lord) Rutherford discovered the behaviour of radioactivity in an element called thorium that disintegrated into a gas, that it became possible to calculate the age of a rock accurately, by measuring the amount of radioactive decay in the rock itself. All radioactivity decays over time, so by measuring this we can calculate the age of any radioactive element, such as a radio-isotope uranium, which produces helium as it decays. The amount of helium in the rock therefore determines the age of the rock itself. The decay is unaffected by chemicals, temperature, pressure or indeed anything. This is by far the most accurate means of ageing many things, including rocks.

Through their study of sedimentary rocks, early geologists estimated that the process of building up the layers of the Earth had taken millions of years. But when they discovered huge folds in the rock, like this one on South Georgia in the southern Atlantic Ocean, they realized that it must have taken much longer.

Rutherford soon surprised the world by finding that uranium-rich material from Connecticut was at least 500 million years old. In an instant he had increased the age of the Earth fivefold. However, this was only one piece of rock and no one was sure whether it was indeed the oldest.

Finding the oldest rock on the Earth

The search for the oldest part of the planet had begun. It wasn't until the 1960s and 1970s, with the help of new isotopic dating techniques, that the true age of the Earth began to be revealed. Stephen Moorbath, working with these techniques at Oxford University, England, heard that some mining works not far from Nuuk on the southwestern coast of Greenland might have one of the oldest rocks in the world. Sure enough, he discovered the rock to be a lot older than anything Rutherford had studied – an amazing 3.75 billion years. No older rock formation has been found anywhere on this planet.

Scientists knew that the oldest rock didn't reflect the true age of the planet, as the Earth was still in great turmoil between the moment of its creation and the time when the first solid, permanent rock was formed. To discover its birth date, they had to look elsewhere. Since the Earth was born when the solar system was formed, an object from outer space that remained intact from the beginning of its creation had to be found.

The only way to find such an object was to visit the huge craters that had been created by meteors from outer space. Meteors are lumps of rocks left over from when our solar system was created – that is, when the Earth was just beginning to be formed. These lumps are continuously moving around the sun just as the planets do, except that occasionally they collide with our planet. In 1999, a place known without coincidence as Meteor Crater in Arizona was investigated and the remaining pieces of a meteor, called meteorites, were found inside the crater. When the latest dating technique was used on these rocks, they were discovered to be 4.55 billion years old. Scientists now agree that this is when the Earth was born.

The Earth is more dynamic than the moon and Mars so, although we have experienced just as many cosmic impacts as they have, the surface of our planet can change quickly, healing over the wounds created by meteors crashing into it. As we have seen, we can find rocks as old as 3.75 billion years old, but this is very rare. Most of the existing surface of the Earth is in fact only 100 million years old, because new landscapes are constantly being formed. With the help of our Time Machine we can start to see nature's forces at work and gain a better understanding of how our planet was formed and why it looks the way it does today.

The Earth's birth

When we set our Time Machine to go back 4.55 billion years, we see that the Earth has not yet been formed. Instead we see a huge cloud of interstellar gas and dust, spread out across something much bigger than our solar system will be by the time of the twenty-first century. The Big Bang – when time and our universe started – has happened 10 billion years earlier. The densest part of this cloud then started to collapse and 99.9 per cent of it, largely hydrogen gas, was pulled together to make the sun. The result of this massive amount of material coming together created thermonuclear reactions in its core where hydrogen atoms fused together to form helium, which in the process released the enormous amounts of energy that come to us on Earth as light.

At the beginning of our solar system as we know it, the remainder of the gas and dust begins to rotate around this new sun and flattens out into a far-flung disc, known as the solar nebula. Everything moves in a similar orbit around the sun. The dust particles begin to clump together to form granules; they in turn form pebbles, which join together to form rocks. The attraction of all this matter is the result of gravitation pull. As each rock grows in size, so does its mass, and accordingly its gravitational pull increases too. Larger rocks clump together into gigantic rocks several kilometres across, and so on until eventually spherical planets begin to form.

An interesting feature of our solar system is the way all the material in it moves in the same direction around the sun. This orbiting helps to keep everything together within the solar system. (The sun itself spins in the same direction, as do all the moons around their parent planet — and they still do today.) The orbit of each of these objects becomes faster as they grow bigger.

The Earth, which at this time has no moon, grows as more and more rocks are drawn to it with increasing gravity. Meteors and other debris bombard the planet for at least 500 million years, producing enough energy and heat to vaporize any water or melt any rock that may be present. At this point, gravity separates heavy elements from lighter ones, with the heaviest elements, such as iron and nickel, sinking to form the Earth's core, while the lighter ones, such as silicon, magnesium and aluminium, gradually rise towards the surface.

As these elements separate, the centre of the Earth grows hot and is heated even further by the decay of radioactive elements in its interior. Gases released from magma deep within the Earth escape through cracks and fissures in the surface and start to form the early

Meteor Crater in Arizona, USA, was created, as its name suggests, by the impact of a massive meteor from outer space. By examining tiny fragments of meteorite, the terrestrial remains of the meteor, scientists were able to calculate the true age of both our planet and our solar system.

The universe began with
the **Big Bang**, which also
marked the beginning of
time, 14 billion years ago.
Ten billion years later, our
solar system began to form.
Looking out into space
at nebulae, such as the
Horsehead Nebula in the
constellation **Orion**, where
there are young stars, we
can study the creation of
other solar systems like ours.

atmosphere. These gases are mostly methane and ammonia, highly toxic to all life forms, but life does not yet exist. As there is little or no free oxygen, there is no protective ozone layer; and so ultraviolet light, also harmful to any new life forms, penetrates the atmosphere at full strength.

Then, around 4.5 billion years ago, a Mars-sized planet hits the Earth with incredible ferocity. It disintegrates on impact and a massive jet of vaporized rock is propelled into space, forming a cloud of molten rock that orbits the Earth and finally condenses into the moon.

After this impact, the Earth's rotation is at first speeded up, producing a day length of only five hours. Later, the tidal forces exerted on the Earth by the moon slow it down again. The moon is about half the distance from the Earth that it is by the time of the twenty-first century and so looks twice the size; as the Earth's rotation slows, it loses its angular momentum, which is the force holding the two together, and the moon starts to drift away by 4 centimetres (1½ inches) every year. Not much, but over billions of years the moon appears smaller and smaller in the sky as it moves away, and will continue to do so for billion of years into the future. The spin of the Earth, whose axis had been upright, is now tilted at 23.5 degrees due to the colossal impact and so seasons are seen for the first time, as the Earth moves around the sun.

For the next several hundred million years, the Earth is very hot, several thousand degrees Celsius, with its surface virtually covered in molten lava produced by thousands of volcanic eruptions. Then slowly, parts of the surface begin to cool and a more solid, harder surface appears. As the meteorite bombardment finally slows down, water condenses in the atmosphere and torrential rainstorms ensue, increasing the cooling effect on the land. Over several million years of continuous rain, oceans begin to fill.

By about 3.9 billion years ago, the Earth's environment has changed from being highly volatile and unstable into a more hospitable

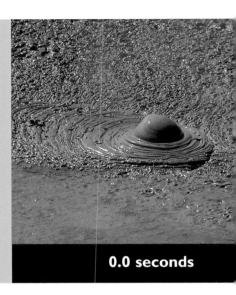

The hot Earth

Hot steam and gases create bubbling mud in Rotorua, New Zealand. It is a reminder that a few kilometres below the Earth's cool, hard crust, the planet is very hot. The exploding mud happens within less than a second, but the heat that it produces is the result of the Earth's creation, 4.55 billion years ago.

0.0 seconds

place. Its surface is completely covered with a hard crust of basalt and silica. Volcanoes are still erupting and lava continues to spill out on the land, but to a far lesser extent. The seas surround the higher parts of the dry barren land. As the sun heats both land and sea, it creates weather systems. Because of the Earth's tilted spin, heat falls unevenly, so weather patterns become dramatic, with more chaotic changes and temperatures rising and falling to a far greater degree than they do in the present day.

At this stage we can look out across our solar system and see other new planets orbiting the sun. There are two types. Close to the sun are the solid balls of rock with a thin atmosphere – Mercury, which is closest to the sun, and then Venus, Earth and Mars. Further away from the sun the planets have a rocky core with an enormous amount of gas around it. These are the giants – Jupiter, Saturn, Uranus and Neptune. The outermost planet, away from the heat of the sun, is a frozen one called Pluto.

Back in the twenty-first century, the Earth spins on its own axis every 23 hours and 56 minutes and moves around the sun every 365.24 days. The moon orbits the Earth every 29.5 Earth days, while the sun spins on its own axis every 25.4 days. As we shall see, the first three of these periodic episodes of spin and rotation have become an essential part of timekeeping not only for ourselves, but also for all life on Earth.

0.5 seconds | **0.7 seconds** | **1 second**

The birth of the planet was extremely violent and dynamic, and the Earth is still reeling with aftershocks. Even today its surface is not stable and from time to time we are reminded of the underlying energy that occasionally erupts to the surface.

The shifting Earth

As the Earth's surface cooled down in the first billion years of its history, a hard outer crust began to form, but instead of one hard shell it is made up of 15 massive rigid plates. These plates still exist today and make up our continents and major oceans, each covering hundreds of thousands of square kilometres. Despite their enormous size, these plates are not fixed, permanent features; from time to time the usually solid and stable ground beneath our feet begins to shake with an almighty force.

Earthquakes are the result of these giant plates moving. Their devastating effects happen when rock literally snaps in two after being bent by the continuous movement of continental plates. Plates rarely move smoothly past each other, but instead are held while pressure builds up. Finally movement and energy are released all at once, causing havoc. Most earthquakes occur several kilometres underground, but the energy released sometimes causes the Earth to shake so violently that buildings collapse, which is the main reason people are killed. In certain parts of the world, tremors are a regular occurrence. Some are so minute you can hardly feel them, and if larger ones are common you soon get used to them and eventually ignore them. That is until a massive one happens.

Earthquakes are measured on the Richter scale from 0 to 9, with each number representing 30 times more energy release than the last. The ones that cause serious damage tend to measure 6 or more. On 26 December 2003, in the city of Bam, Iran, an earthquake of a magnitude of 6.6 was destructive enough to kill over 30,000 people and injure as many again, levelling most of the city within minutes. Fortunately, the majority of us will never witness such an event unless it is recorded on film. In Kobe, Japan, on 1 January 1995 television cameras inside office buildings recorded the great Hanshin–Awaji earthquake. The pictures showed desks and blinds being shaken with great ferocity. The might of the earthquake left the city in ruins, with collapsed elevated highways and demolished buildings. Dust and smoke hung in the early morning air and thousands of people died. For several days the Earth rumbled with aftershocks as the crust settled down following this catastrophe.

The problem is that earthquakes can occur in an instant, without any warning, and with awesome and overwhelming power. We can predict that an earthquake is going to happen, but not exactly when. If only people had a Time Machine to see what was going on.

Only 150 years ago a huge earthquake rocked the township of Wellington, the capital of New Zealand, destroying most of its buildings. On Tuesday 23 January 1855 at 9.17 p.m., a massive 8-plus earthquake shook the city, bringing carnage in its wake. Despite the huge collapse of buildings there were only a few fatalities, no more than ten.

Even before the great shock had died away, the high tide at Lambton Harbour had risen by 2.5 metres (over 8 feet) and flooded shops and houses along the shore. The water receded almost immediately to several metres lower than the lowest tide had ever been. For the next eight hours the tide rose and fell every 20–25 minutes, rising 2–3 metres (6½–10 feet) and receding again to a little over a metre (3 feet) lower than at spring tide. But something far more fundamental had happened to the sea level, and that would be revealed the next morning.

As day broke, clouds of dust rose from the rubble and people were seen in their doorways or sitting out in their gardens. The first thing anyone noticed was the extremely low tide. The beach at Lambton Harbour extended far beyond its usual limits. This was later found to be the result of a massive uplift of land that raised it over a metre (3 feet).

Large holes appeared where the fault lines had shifted, creating gaps 2.5–2.7 metres (8–9 feet) wide. The land had not only shifted north and south along the fault line, it was also twisted upwards, causing the land to rise and creating the lower sea levels in those areas.

The road running along the shore between Wellington and Lower Hutt, a town about 15 kilometres (10 miles) north, which could previously be passed only when the tide was low because the sea could come right up to the cliff edge, was now high and dry. From that day onwards, the road has been passable at any time despite the tides. Today it serves as the main motorway between the two

A haunting image captures the devastating aftermath of the Kobe earthquake in Japan in January 1995. Without any warning and within less than a minute, the city had been completely destroyed. Events such as these wake us up to the fact that our planet is still reeling from its birth, 4.55 billion years ago.

The main highway between Lower Hutt and Wellington on New Zealand's North Island was built on a raised beach that only appeared after the 1855 earthquake.

Within less than a minute, the whole land was lifted up out of the sea by a few metres. Previously the road could only be used when the tide was low.

cities, even though it is built on the fault line that runs through the southern tip of the North Island.

The shallow waters between the mainland and the peninsula of Miramar also rose dramatically out of the sea to form a natural bridge. Today Wellington's international airport stands on some of the land that only 160 years ago was below sea level, thanks to the power of the earthquake.

At Turakirae Head, just a few kilometres east of Wellington, there is a beach that clearly shows four great uplifts. The lower line, parallel to the shore, was the beachhead before the 1855 earthquake, the next above this line is the 1460 earthquake, and the other two above that reflect changes of about 3000 and 4900 years ago.

It is now known that a few seconds before 9.15 p.m. on Sunday 21 January 1855, at a depth of 25 kilometres (15 miles) below Cook Strait, between the North and South Islands, some 40 kilometres (25 miles) southwest of Wellington, a large section of the Earth's crust suddenly ruptured, releasing energy a thousand times more powerful than the Hiroshima atomic bomb. The shock wave radiated outwards and upwards at a speed of 6 kilometres (4 miles) per second.

So what about the future? In our Time Machine, we can see that the coastline will lift up again in a couple of hundred years' time. In 35,000 years, assuming it is raised by just 2 metres (6½ feet) every time there is an earth-quake of a similar magnitude to the one in 1855, it could be 200 metres (650 feet) above sea level. Where would Wellington be then?

If the people of New Zealand had a Time Machine and could see exactly what dramatic shifts of the landscape can happen over eons of time, they would never live or build near places like Wellington. The next major earthquake may not be for another 250 or 300 years, but we just do not know, as the tragedy at Bam clearly showed. The peaceful period, when nothing more than the odd rumble is happening, gives us a false sense of security. We put an awful trust in nature, until the next tragedy.

When the heat reaches the surface

The volcanic White Island lies like a speck on the horizon in the Bay of Plenty, off the North Island of New Zealand. Gradually it begins to grow, as the helicopter approaches it. The best way to see this island is from the air, but even so it shows only one-third of its true height, as most of it is under the sea. Its size is deceptive from the air and is only really appreciated when you land. Standing in the middle of a volcano is an awesome experience. The white and grey ashen moonscape island is stark. Towering

around you are the 100-metre (330-foot) high walls of the main crater. A short walk takes you to the edge of another inner crater. Below your feet there is a sharp drop of some 80 metres (260 feet) down to a lemon-coloured lake of almost concentrated, bubbling sulphuric acid, its surface covered with pungent steam.

Little survives here; there are just a few trees on the west slope that faces towards the sea. Moist and occasionally wet winds give a gentle hand to their existence. The stench of sulphur vapour and hydrogen sulphide swirls around and almost encloses you. Dotted on the edge of the main crater wall, steaming hot vents act like pressure valves, letting built-up steam and gases escape with a ferocious hiss. These are fumaroles, blasting

New Zealand's volcanic White Island looks quiet, but underneath the surface pressure is building. Today scientists can predict eruptions within a few hours, **allowing visitors enough time to leave. It was a different story in 79 AD, however, when the people of Pompeii, Italy, had no warning of Vesuvius' eruption.**

choking acidic gases from yellow pores the size of rabbit holes, gateways to a fiery hell below. The yellow is from the delicate-looking crystals of pure sulphur that materialize around the vent as the gases cool.

The relentless sound of a jet engine roaring from these vents is a constant reminder that the volcano is very much alive. It just happens to be a little quiet – for the moment. Scientists from the Geological and Nuclear

Sciences (GNS) continuously monitor the slightest changes in the volcano's structure. They know that the surface is swelling from the huge pressure building up below. Every day a party of 100–200 tourists arrives by boat to explore this extraordinary uninhabited island. They don hard yellow hats and gas masks, just in case; little protection if the island decides to have a good blow, as it did relatively recently.

Back in 1972, the ground on the island started to rise and bulge as much as 10 centimetres (4 inches) over four years. Then, without warning, in December 1976, the volcano erupted. For the next ten years it continued to pour ash and smoke up several thousand metres (tens of thousands of feet) in the air. Then it all went quiet again.

However, since 1992, another measurable rise in the ground has started and, in 1995 when it started to quicken, the public were warned about it. Despite this, people continue to come to the island and everyone is still waiting.

When enough pressure builds up beneath the ground, from several kilometres below the surface, hot molten lava is shot to the top. It is very much like shaking a fizzy drink – the pressure builds up so much that the Earth literally explodes. Today scientists can predict if a volcano is about to erupt by listening out for certain sounds underground as well as measuring the seismic activity – the rumbles that are happening below. The rate at which the ground vibrates gives a clue as to whether it is going to blow.

In 1996, when Mount Ruapehu on the North Island of New Zealand erupted, scientists from GNS were able to monitor the events that led up to a major explosion. Using sensitive chemical analysis and seismic recordings of the existing crater's lake, they detected rises in temperature, chemical changes and a lot of rumbling going on deep below. The dominant vibration was measuring around 6–9 vibrations a second (6–9Hz). But then at 3 p.m. on Saturday 15 June a sudden change occurred: vibrations dropped down to 2Hz, indicating that the molten lava had started to flow deep below the volcano. At 5 p.m. the GNS made a public statement warning people of a pending eruption, especially the skiers on a resort next to the mountain. They knew it was about to happen, but not exactly when.

0 seconds

20 seconds

It happened 37½ hours later, on Monday 17 June at 7.30 a.m. Ash and rocks were flung in all directions, up to 2–3 kilometres (1½–2 miles) away. The crater's lake was emptied, but ever since it has been filling, and today scientists are worried its walls might collapse, causing a mud flow down the mountainside and bringing havoc in its wake. But close geological monitoring and public warnings should prevent any human tragedy in the future.

Volcanic eruptions can change the landscape dramatically in a very short time. When Mount St Helens in Washington State blew its top off in 1980, causing one of the biggest eruptions in recent history, the whole side of the mountain was thrown for several kilometres around. Forest was devastated, with trees completely flattened by the blast.

In the last century, an estimated 100,000 people were killed as a result of eruptions around the world, most of them caught off guard by the volcanoes' unpredictable nature and suffocated by the poison gases it emitted.

By travelling back in time in the Time Machine we can see how quickly the whole process happens. In 79 AD, when Mount Vesuvius near Naples in Italy erupted, the city of Pompeii had a sudden shock. All the residents were killed, mostly by suffocation – a total of 3360 deaths. Within hours of the eruption, a pyroclastic flow of fine dust and poisonous gases came rolling down the mountain, hugging the ground, at over 160 kilometres (100 miles) per hour. It was the volcanic ashes that finally buried and preserved the city and its inhabitants. There they remained until 1748, when an exploration to see what had happened found their bodies in the positions in which they had died nearly 2000 years earlier.

Volcanoes are found where the Earth's crust has become thin due to tectonic plates moving under pressure. They are usually close to a point where one plate meets another and one of them is pushed several kilometres down into the magma, a process known as subduction. Such volcanoes are found in the famous 'ring of fire', around the edge of the Pacific Ocean – from the Andes in South America, up along the west coast of North America, from Alaska down through the Aleutian chain and Japan, Indonesia and New Zealand. Europe's

30 seconds

10 minutes

A question of timing

This eruption of Mount Ruapehu in New Zealand took place at about 5 p.m. on 23 September 1995. A skier who was close by took these photographs just as it started to erupt. The explosion took place over a 10-minute period; however, further minor eruptions, depositing ash on the upper slopes, lasted for days.

only volcanoes are found in the south, where Mount Etna in Sicily erupts every few years and Mount Vesuvius is now the only active volcano on the Italian mainland.

On the surface, the work of volcanoes is hardly seen for most of the time, but far beneath the Earth's crust pressure can build up quickly. Submerging plates pushed along fault lines into the crust usually carry water with them. As this is heated, the molten rock or magma is forced up through cracks in the crust to erupt, building a volcanic mountain.

When a volcano explodes, magma from below the earth's crust is quickly released. As it froths, it turns to lava and volcanic gases. If the gases are volatile, the lava explodes, flinging huge rocks out of the crater. Clouds of smoke and ash puff and bellow out of the volcano, while lava usually oozes and spills from the crater and surrounding cracks down the mountain's slopes.

Magma can be either sticky or runny. Sticky lava flows slowly and can be explosive, while runny lava flows, like a river, quickly and further, and it is less explosive. The sticky sort creates steep-sided cone-shaped volcanoes, while the runny kind produces more gently sloping, flattened shapes. The hotter the lava is, the more it flows, and temperatures as high as 1200°C (2100°F) have been recorded.

The Earth's tallest single mountain

The snow-peaked Kilimanjaro stands nobly above the hot, dry grasslands of Kenya. Recognizable everywhere, it is just 3 degrees south of the Equator, close to the border with Tanzania. Its distinctive cone shape and white cap give us a clue as to why it exists in such a hot country. It is both the largest volcano and the largest free-standing mountain in the world. So how did it achieve this accolade?

We can find out by using our Time Machine, setting our course to just over two million years ago, when there were just rolling plains where Kilimanjaro now stands. Speeding forward in time, we see that the formation of the Great Rift Valley, which now stretches 10,000 kilometres (over 6000 miles) from Lebanon to Mozambique, had just begun, when two plates moved apart and the plains twisted and slumped. Fissures and cracks formed and lava began to flow out, forming a string of volcanoes along a 100-kilometre (60-mile) gash in the Earth's crust.

Three-quarters of a million years ago, just 80 kilometres (50 miles) east of the emerging Great Rift Valley, Kilimanjaro started off its life out of three vents, Kibo, Mawenzi and Shira, which grew slowly, forming three steep-sided cones. In a bizarre way it is as if they were taking part in a race to see who would be the tallest. By half a million years ago, they had built up to a height of 5000 metres (16,400 feet). Then the momentum was stalled when Shira collapsed in on itself and became inactive. The collapsed crater, the caldera, is still visible today as a broad plateau. Both Kibo and Mawenzi continued to explode and grow, reaching 5500 metres (18,000 feet). Then Mawenzi gave up and became dormant. Surprisingly, it quickly eroded away, forming steep rocky cliffs, the remains of the hardened plug from which magma once flowed. All signs of its crater disappeared. However, Kibo continued to grow and became very active 350,000 years ago, growing another 400 metres (1300 feet) to reach a staggering 5900 metres (19,350 feet). This height was only achieved in recent years, as Kibo is thought to have been an active volcano until just a few hundred years ago.

Today the white-capped summit of what is now known as Mount Kilimanjaro has a crater 2.3 kilometres (1½ miles) wide, and inside that is a smaller crater 350 metres (1150 feet) wide. Here the sulphur steam vents are still vigorous, a constant reminder of the mountain's origins and its newness. No one knows whether it will ever erupt again, but one thing is for sure. In the race to the top Kibo finally won out. And the secret of Kilimanjaro's success is that three volcanoes joined together to make the tallest volcanic mountain in the world.

The snow-capped peak of Mount Kilimanjaro in equatorial East Africa took just 2 million years to become the tallest single mountain in the world.

This is very quick in geological time, when one considers that the surrounding wildlife has changed very little over the same period of time.

Massive changes

The two-million-year timescale in which Kilimanjaro was created is tiny in comparison to that of other geological features on the planet – some take place over ten times as many years. Most of these changes are incredibly small over the course of our lifetime. But by using the Time Machine, we can add up all these small changes and view the great expanse of time that was needed to create some of the greatest natural wonders of our world.

Cracks in the Earth

Under the hot blistering sun of the dry and arid countryside of Djibouti, in northeastern Africa, workmen are repairing cracks in the paved road. It is a task they have to do every year at the same spot. In fact, it is not just these men who have a job for life, but countless generations to follow. The force that is causing the cracks to appear is monstrous and relentless. It will go on not just for a few years, but for thousands of years to come.

Amazingly, 800 earthquakes a day shake this region. The cracking starts in the Red Sea and runs south for 5600 kilometres (3500 miles) to Mozambique, forming the famous Great Rift Valley, home to so much of East Africa's wildlife. The Earth's crust is thinning because it is stretching as two continental plates begin to move away from each other. The greatest thinning between the two plates has already caused the land to collapse and created the Red Sea, forming a deep-water channel between the Horn of Africa and Arabia. Now a similar thing is happening on land in the adjacent Afar Triangle in Djibouti. For the last 25–30 million years, the Triangle has seethed with volcanism. The Red Sea and the Gulf of Aden are active and spreading centres of activity. Magma, fuelled by the extremely hot molten lava welling up in the asthenosphere, is forced up beneath the crust. When it reaches the surface it cools and hardens at the parting edges of the plates beneath the Red Sea. On land the magma is forming domes just beneath the surface. Here faults fracture the crust and segmented blocks of the rift tilt in opposite directions, causing deep depressions in the ground.

Some 1100 kilometres (700 miles) southwest of Djibouti lies Lake Turkana, created because the thinning crust has caused splitting and rifting. Water running off the land has filled the collapsed and dented surface, producing this huge lake. Further south again, an even deeper cigar-shaped trough has become Lake Tanganyika. Here, the rifting appears to be happening more quickly but over a longer period of time – sediments extend 5000 metres (16,500 feet) deep. At Lake Malawi, another 250 kilometres (150 miles) to the south, there is only a little stretching in the Earth's crust. This is probably the youngest part of the Great Rift. Seen from space, the whole of the valley seems to be unzipping itself

East Africa's future

The Great Rift Valley is the beginning of a massive land split that will cause East Africa to separate from the rest of the continent in about 50,000 years time. In 5 million years, East Africa will be a totally separate island, well distanced from the mainland, just as Madagascar is today.

Today

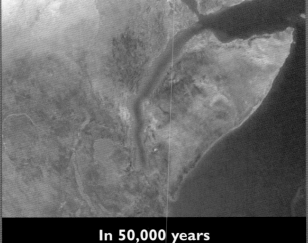

In 50,000 years

from the north down to the south, driven by huge underground forces.

So what can the Time Machine help us find out about the future for East Africa? Well, no one is sure if the continental plates will continue to split apart but if they do, at some point, perhaps 50,000 years in the future, the sea will break through from the north into the Afar Depression and the valley floor will drop further down. We will see the sea beginning to creep down the valley, turning the great wildlife sanctuaries of the Masai Mara and Serengeti into a tropical sea. Lake Tanganyika and Lake Malawi will become deeper and fill up with the advancing sea. East Africa will split off into the Indian Ocean, just as Madagascar did 165 million years ago, and, like Madagascar, will develop its own unique wildlife.

In Iceland, cracks of greater magnitude can also be seen; the island is widening by 20 centimetres (8 inches) a year – that is faster than your fingernails are growing. Iceland is one of the most active islands on Earth. One third of all the lava that has erupted on land over the last 500 years has been here, pushed up from just a few kilometres below the surface. The lava is mainly located at the centre of the island, in a crack called the Mid-Atlantic Ridge, which divides the country and to which Iceland owes its existence. The ridge continues southwards right through the Atlantic Ocean – from Iceland to the Falkland Islands lava pours out of its huge

chimney stacks. It is responsible for moving not only North America and Europe apart, but South America and Africa too. All of this has happened within 335 million years. The sheer scale of what is happening can be appreciated more fully when seen over time in our Time Machine.

When the first deep-sea submarine dives went down to see the Mid-Atlantic Ridge they soon confirmed that it was the site of intense volcanic activity. There was evidence of young lava flows everywhere, oozing lava like toothpaste from a tube. The phenomenon known as black smokers is present here, producing hot black clouds of 350°C (660°F), a superheated plume of muddy water that pours out of the seafloor. These black smokers are caused by the flow of sea water through the oceanic crust, which can be happening as much as 1500 kilometres (1000 miles) away from the ridge. Cold sea water penetrates cracks in the crust and is superheated deep below. It then rises rapidly to escape through the seafloor. In just a few million years the entire ocean is pumped through cracks like this. Salt is extracted and new minerals added to the gushing water. These chemical exchanges balance the entry of salts and ions brought into the oceans by rivers, preventing the seas from becoming a lethal chemical soup, too deadly for life to survive.

Extraordinarily, life does survive in the most hostile conditions around the base of these hot plumes. Species

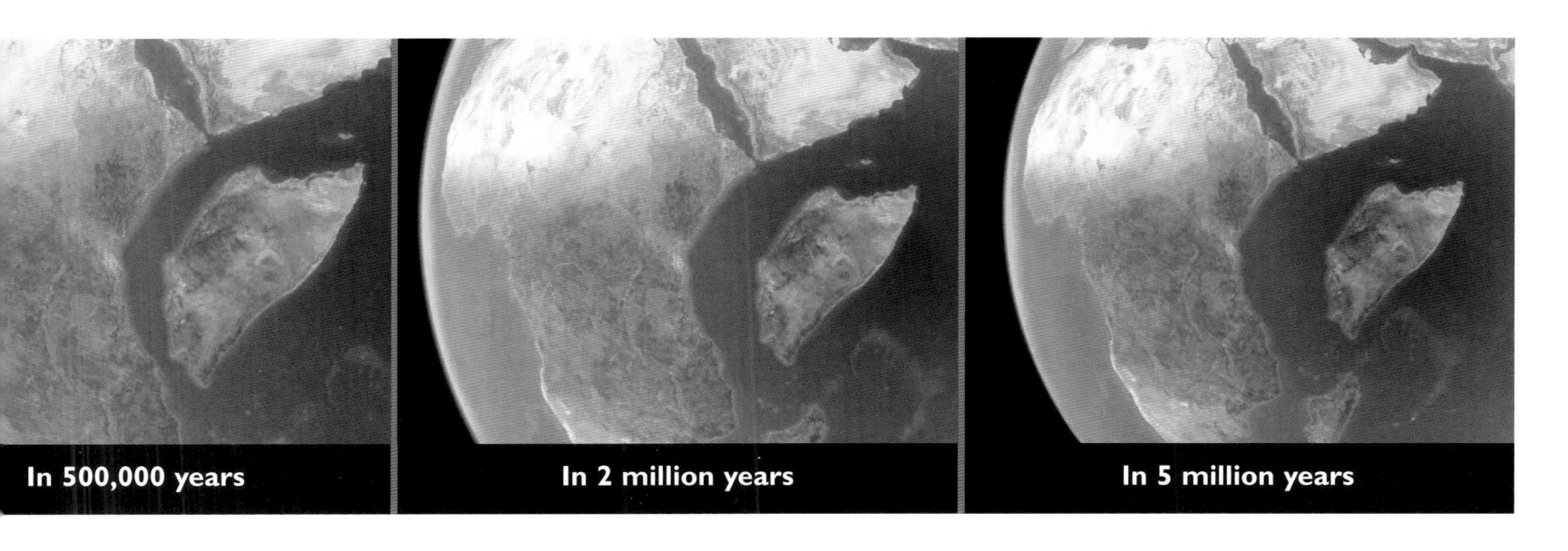

In 500,000 years | In 2 million years | In 5 million years

Rise of the highest

The Himalayas are the result of a gigantic collision of continents. The rise of this mountain range has been faster than the powerful erosive weather wearing it down. It still took plenty of time to achieve its current height though, about a million times the length of an average human lifetime.

50 million years ago

40 million years ago

of clams, mussels, shrimps and tube worms, many of them new to science, have adapted to survive without relying on the sun as a source of fuel to sustain them.

The rooftop of the world

Eight thousand metres (over 26,000 feet) above sea level, high in the Himalayas in Nepal, you can find marine animals fossilized in the rock. The question is, how did they get to the top of the highest mountain range in the world? The Time Machine is one way of discovering the answer.

When these fossils and the sediments around them were dated, they were discovered to be 60 million years old. Setting our time and course in the Time Machine we can fly back to the past. We discover that this marine animal was fossilized deep in the Tethys Ocean, which lies between the independent subcontinent of India and the rest of Asia. India has long separated from the super-continent of Gondwanaland, but is moving steadily north towards Asia. Speeding through time it seems as if India is drifting on the sea, but in fact this huge landmass and its plate are moving on a hot molten bed of magma. Finally 55 million years ago, when the continent crashes into Asia, the bed of the fast-disappearing Tethys Ocean is pushed up along the faults. The Indian plate will travel a further 2000 kilometres (1250 miles) north as it plunges under the growing Himalayas.

At first small folds appear, all in a series parallel to each other, but at right angles to the advancing continent. The colliding subcontinent is moving so fast that the folding of the Asian crust is enormous, creating a vast mountain range. The rocks here are being squeezed and thickened, compressed and crumpled in very complex ways. As the advance continues, the folds get bigger and the mountains higher. The growth is so rapid that even powerful erosion has little effect on it. Because of the enormous pressures involved, hardened metamorphic rocks like granite deep in the continent's crust are pushing up along many faults with incursions of molten rock. Eventually the mountains reach their maximum height and begin to get wider rather than higher, forming the great Tibetan plateau.

Today the plains of India are still sliding under the Himalayas along a huge fault line that runs along the southern edge of its foothills. As you fly north over it you see range after range of mountains, each one higher than the last, until you come to Mount Everest. This is where the folded sediments of the old Tethys Ocean have ended up, along with their marine fossils.

If we now fast-forward into the future we will see that the Himalayas are not getting any taller, as they will be eroded as fast as they rise, and India's push into Asia is slowing down. With the enormous weight of this gigantic mountain range on top of a thin crust, the Himalayas will quickly collapse and erode away. Eventually, some

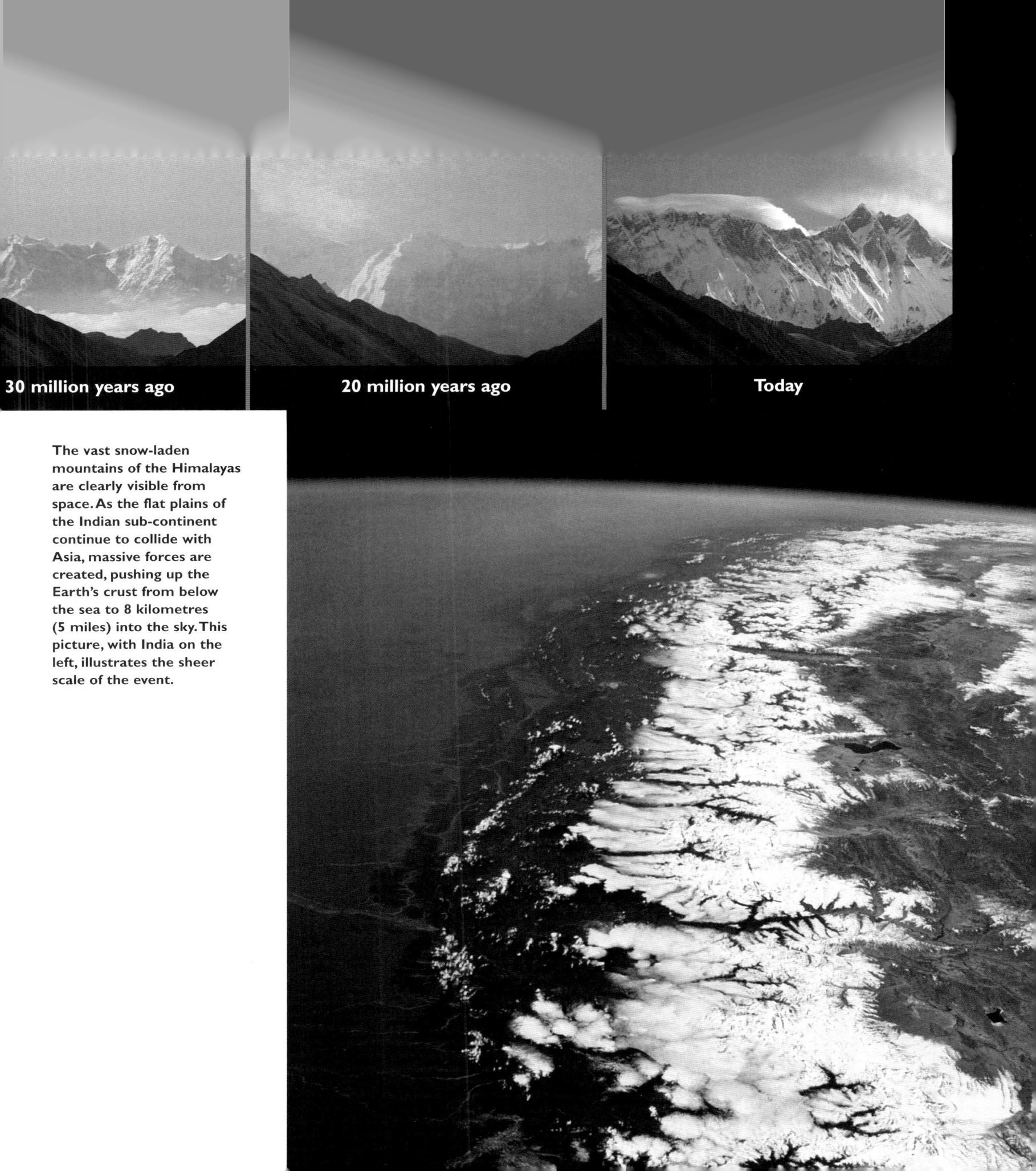

30 million years ago

20 million years ago

Today

The vast snow-laden mountains of the Himalayas are clearly visible from space. As the flat plains of the Indian sub-continent continue to collide with Asia, massive forces are created, pushing up the Earth's crust from below the sea to 8 kilometres (5 miles) into the sky. This picture, with India on the left, illustrates the sheer scale of the event.

geologists believe, their disintegration will be so complete that the land will fall below sea level and create a new sea or ocean. The legacy of Mount Everest will be to return that little marine fossil to the sea in which it once lived, if it does not disintegrate on its journey.

Eroding mountains

So how powerful can the force of erosion be on a mountain? It is hard to believe that the peaks of the Himalayas, the European Alps or the Rocky Mountains could be much, much higher than they are. But they could be between three and five times their current height if it were not for erosion continually wearing them down.

One of the biggest measured erosions has happened in the Southern Alps of New Zealand, which have been pushed up by the collision of the Australian and Pacific plates over the last 25 million years. The Earth's crust has been compressed along a fault line that has been squeezing it tighter and pushing this mountain range above sea level. These mountains are among the fastest growing in the world – if no erosion had occurred, they would be 20 kilometres (12½ miles) high, reaching way beyond the stratosphere. However, erosion has kept them below 4000 metres (13,000 feet).

The effect of weathering on rock can be either mechanical or chemical. Mechanically, heating by day and cooling by night can cause the rock to crack. Then water penetrates and freezing causes it to expand, shattering the rock into pieces. Over time, we can see the rock falling to form slopes of loose scree at the bottom of rock faces. Frost heave can also loosen chunks. As soil freezes, crystals grow vertically, lifting the particles up off the surface. When the ice melts, gravity pulls the loose rock down.

Chemically, mountains can be eroded when water and oxygen combine with certain minerals to form oxides and clays. Over time, these can have a severe corrosive

effect, weakening the rocks and causing them to break up, especially in warm and wet places like the Southern Alps. Interestingly, trees and other plants can protect a mountain from erosion, effectively wrapping it up against the harsh elements of water, wind and ice.

The most dramatic way of removing huge amounts of rock from a mountain is an avalanche, which can reduce the size of a mountain within minutes.

One of the most spectacular avalanches in living memory happened shortly after midnight on 14 December 1991. Climbers camping beneath Mount Cook, the highest peak in the Southern Alps, woke to a loud rumble that turned into a roar. They saw sparks as rocks collided together. The noise continued throughout the night, and in the morning a cloud of grey dust hung over the valley. A massive part of this famous peak had completely vanished. The tallest mountain summit in New Zealand had just been reduced by 10 metres (33 feet) to 3754 metres (12,313 feet), with 14 million cubic metres (nearly 500 million cubic feet) of rock scattered below it. The avalanche descended into the valley at speeds of up to 400 kilometres (250 miles) an hour. The momentum was so great that it spread across the full width of the valley and delivered debris 6 kilometres (4 miles) beyond the summit.

The most extraordinary thing about all this is that it happened so quickly and was unaided by earthquakes, which are common in this part of the world. The main avalanche lasted just under a minute, but its power was so great that the shock waves registered on some seismographs, the instruments that measure earthquakes.

What the future holds

If we travel forward 50 million years in our Time Machine we will find that South America has split off from North America. Baja California and the area west of the San Andreas fault line have split from the rest of North America and are much closer to Alaska. Africa is further north, too, and is much closer to Europe, with the Mediterranean Sea reduced in size. The Himalayas are eroding fast and have lost much of their magnificence. Australia is further north, off the coast of China and much closer to Japan.

The San Andreas fault runs south of San Francisco in California, dividing two continental plates – the North Pacific and North America. Every time there **is an earthquake, the North Pacific plate moves a few centimetres north, over the North America plate. Over millions of years it can move hundreds of kilometres.**

The power of the elements

Most of the sculpturing our planet undergoes is a result of the climate and weather systems that are powered by the sun. However, the timescales involved in the erosion of rocks are baffling and often beyond our comprehension. The full magnitude of the changes can only be revealed when we travel back through time at great speed.

The Grand Canyon

One thing that makes our Earth unique is water. Water covers two-thirds of the planet but it doesn't just sit there – it is influenced by the sun and the atmosphere and can be a dynamic force for change. The seas around the Equator are heated by the greater strength of the sun – water evaporates into the warm atmosphere and is then carried several if not hundreds or thousands of kilometres by the prevailing winds and jet streams that operate in the higher atmosphere above the Earth. When this warm, moist air hits cold air, as it does when it is lifted high over mountains into the upper atmosphere, it cools and the evaporated water particles combine together and fall as rain droplets. Rainwater is pulled towards the Earth by gravity and when it hits the ground it is pulled further by the same force, down the mountains, accumulating in streams. The streams join other streams and soon a river is formed. The rivers flow down the mountains to the lowest point at their base, carving out valleys as they find their way eventually back to the sea.

Nowhere is the force of water seen more dramatically than the Grand Canyon in Arizona. Here the Colorado River cuts into the Earth, shaping one of the most magnificent gorges in the world. This would not be surprising except for one thing – it is in a particularly arid part of the western United States.

The Grand Canyon, USA, is the world's largest ravine, and cuts deeply into the Earth's crust. Sudden but infrequent rainstorms in the desert create flash floods. These carry a mixture of water and rock debris, which acts as a powerful erosive force, down through the valley, carving out steep-sided canyons.

Carving the canyon

The infrequent flash floods in the Arizona Desert are violent episodes. The force of water and debris in a flood cuts into the rock by just a few millimetres, but seen over millions of years, the effect is far greater.

5 million years ago

3 million years ago

If you are lucky enough to fly over the Grand Canyon you are bound to be impressed by the way its steep cliff edges cut deep into the Earth. Looking down perhaps 1.5 kilometres (1 mile) from the rim of the canyon is an awesome experience. How can a river create a canyon of such magnitude?

Although the Grand Canyon is formed by water cutting and eroding through the various rock layers that are being lifted up from beneath the Earth's surface, the Colorado River causes little erosion for most of the year. It is another force that really carves this awesome canyon.

On Tuesday 12 August 1997, this force made itself felt in the Antelope Canyon, a side canyon of the Grand Canyon itself. Known as a slot canyon because of its deep narrow sides, the Antelope attracts many visitors because of its sheer beauty. In some parts there is no water or river at all. People can walk on the sandy bed and admire the intricately carved walls. On the day in question 12 visitors went on an expedition there, but unfortunately they had not checked the weather forecast.

Some 14 kilometres (10 miles) upstream, a powerful thunderstorm erupted with almost 4 centimetres

Antelope Canyon is a side canyon of the Grand Canyon. For most of the year it remains dry, but all that changes when a flash flood started miles upstream sweeps through the canyon, finely sculpting its rock walls. The sandy floor of the canyon can rise or fall by as much as 6 metres (20 feet) after just one flooding.

2 million years ago | **1 million years ago** | **Today**

(1½ inches) of rain falling, half of it in just 15 minutes. At Antelope Canyon there was only a trace of rain, but the thunderstorm had released so much water from the sky that it created a flash flood. The water arrived at Antelope Canyon just 30 minutes later, with devastating results. By the time it reached the visitors it was a 3-metre (10-foot) wave, powerful enough to sweep 11 of the 12 visitors to their deaths.

Flash floods like this occur when excessive water fills normally dry riverbeds, some of which are found alongside currently flowing streams and rivers, causing water levels to rise rapidly in a very short time. Since the soil in the Grand Canyon is baked hard by the sun it cannot readily absorb water. The fact that on the rare occasions when it does rain the water tends to come down in torrents only adds to the problem. The plants that grow here have very shallow root systems so that they can grab as much water as possible when the opportunity arises. Unfortunately these root systems are too fragile to hold the soil in place, so they do nothing to prevent erosion. With lots of water and nowhere for it to go but down to the Colorado River, and nothing holding the soil and rock in place, flash floods frequently roar down side canyons at speeds that can move boulders the size of buses and even small houses. They are more like fast-flowing concrete than water and can be very dangerous. The river is laden with 500,000 tonnes of debris, which adds to its erosive impact, sculpting the canyon wider and deeper. So is this process responsible for the creation of the Grand Canyon?

A popular theory today is that two different drainage basins, or river systems, once existed on opposite sides of the Kaibab Plateau, part of the Colorado Plateau. Many geologists believe that the river system to the west 'captured' the river system to the east through a process called headward erosion, which occurs as the upper part of a river valley cuts back toward its source. Geologists speculate that about 15 million years ago headward erosion began to cut the valley of the western river upstream and east into the Kaibab Plateau. About 5 million years ago that river broke through the Kaibab Plateau to the eastern river basin. The eastern river was 'captured' and began to flow west, into the western river, instead of continuing south in its old channel. The two river systems joined to form the mighty Colorado River, which then had enough power to cut through the plateau even faster, forming the Grand Canyon. (The Kaibab Plateau is almost 3000 metres/9000 feet high and was still rising when the Colorado River cut through it like a cheese wire.) Water from the Rockies had to head to the lowest ground and out towards the sea. It was like a continuous conveyor belt moving vast quantities of sand and mud.

Each individual canyon starts with a surface fissure or crack in the sandstone block. Rainwater runs off the fissure and once it finds a downward slope it becomes a drainage channel. The great erosive power of fast-flowing water then interacts with the soft sandstone, creating a slot canyon and carving and polishing its walls.

All this occurs over eons of time, but in our Time Machine we can see these flash floods shaping the canyon, as the river cuts through the land like a knife through a cake. In one year there are only about two or three of them, but they are enough to take on average about 2.6 millimetres (1/10th inch) off the canyon every ten years. In a million years, this adds up to 260 metres (853 feet), and in 6 million years it is 1.6 kilometres (1 mile), which is the canyon's depth today. Of course, this is only an average; there have been lots of fluctuations through time. In the earliest part of the canyon's formation, erosion is thought to have been much quicker and in more recent times it has slowed down.

The Grand Canyon also owes its unique shape to the different rock layers in its walls. Each reacts to erosion in a different way: some form cliffs, others slopes; some erode more quickly than others. The vivid colours of these layers are due mainly to the presence of small amounts of various minerals. Most contain iron, which creates shades of yellow, red and green.

The next most powerful force is ice. In the colder months, especially on the north rim, water seeps into cracks between the rocks. Earth murmurs or earthquakes may cause these cracks; so too may the constant soaking and drying of the rocks. When the water freezes it expands and pushes the rocks apart, widening the cracks. Eventually rocks near the rim are shoved off the edge and fall into the side canyons, hitting other rocks. On occasions, the fall of a single large rock will cause an avalanche, drastically altering the landscape in the side canyon. The debris from rock falls soon piles up at the bottom and is swept down to the Colorado River the next time there is a flash flood.

So what kind of future is there for the Grand Canyon? In our Time Machine we can see that it will get a little deeper but a lot wider. In places it will become several kilometres wider and several hundred metres deeper as the land continues to rise. Eventually the sides will break down and be flattened. If the climate changes

before this and there is more rain, then the whole levelling process will be far greater.

Rivers of ice

When freezing water turns to crystals it takes on new powers that are extremely destructive to mountains and valleys. On the one hand, water can penetrate rock, freeze and then shatter it. On the other hand, snow can create glaciers – flowing masses of highly compressed ice, which over time are brutal in their power of destruction.

On a clear sunny day, one of the most beautiful scenes in nature is a white, glistening mountaintop against a blue sky. Many people are drawn to glaciers because of their sheer beauty and mystery; some like to hike and climb over them to face the challenges ice can pose; others fly over the glacier by helicopter and land to explore every nook and crevice. Once there, you can feel the sheer power of the ice, sense the awesome strength it has to carve and shape the land. If you listen very carefully, you may hear it groaning. This is a sign that it is actually moving. Those who visit on a regular basis will notice that changes to the glacial features happen almost daily. Cracks open, crevices widen and some parts disappear altogether.

Glaciers are one of the few geological features where change occurs over relatively short periods. But the actual flow can be seen only over hundreds of years, or if we crush the time into a few seconds or minutes in our Time Machine.

The Franz Josef Glacier, on New Zealand's South Island, is a mere dwarf of what it was during the last ice age, when the snow and ice sheet over the Southern Alps was far greater, covering the whole mountain range and beyond. Today the glacier is 13 kilometres (8 miles) long, descending from 2500 metres (8200 feet) to 500 metres (1640 feet). Its steep decline means that it remains intact

The Franz Josef Glacier is the largest of its kind in the Southern Alps of New Zealand. This massive river of ice moves slowly down the valley by, on average, 1.5 metres (5 feet) every single day. However, records show that it has been known to move as much as 7.9 metres (23 feet) in 24 hours.

much further down the warm valley than it ought to, passing through temperate rainforest, so making it one of the most unusual glaciers in the world. It forms at the higher elevations, where warm, moist, southwesterly winds climb the mountains and then cool rapidly, causing heavy snow flows. The snowfalls and rainfalls at these upper levels are quadruple those on the coastline a mere 12 kilometres (8 miles) away.

Gentle snowflakes accumulate on the upper reaches of the glacier, called the neve, and lie there for a while, but they soon move down and melt, not into water but into a whitish granular snow called a firn. This then transforms into a much harder, more solid form of ice, which flows outwards under the pressure of its own weight. The main chunk of a glacier flows in a plastic, almost fluid manner, but its surface layers are always brittle. As it moves over an uneven piece of ground or a steep slope, the ice begins to snap and crack, forming wide crevices. Over several years, the snow mixes with

water and air is expelled to form bluish glacial ice. As it slowly makes its way down the mountain, it is squeezed through the valley, carving it as it goes. Some ice is lost through evaporation or melting, but above in the neve, the glacier is continually renewing itself with new snow and ice pushing the compressed ice further down the mountainside.

In our Time Machine, we can see the glacier move at different speeds. In the future we will see it slipping down the valley like melting ice cream, except at the bottom, where it turns to fast-flowing water. The Franz Josef is one of the most rapidly flowing glaciers in the world, but how did we discover this?

Glacier ice is different from any other ice, largely because it is under so much pressure. In the centre of the main glacier it is clean and white and flows more quickly and gracefully than anywhere else. At the edge, however, are dark strips called lateral moraines, where rock and debris are mixed with the ice. This happens

RIGHT: **The terminal moraine in the foreground is an ancient band of glacial rock deposit, now covered in trees, that was left during the last ice age, some 11,000 years ago, and 10 kilometres (6 miles) from where the Franz Josef Glacier, seen in the distance, has retreated today.**

LEFT: **In the upper reaches of the Southern Alps, soft snow is gradually turned into brittle, hard ice, which is pulled down into the valley under its own weight. It then becomes one of the most powerful earthmovers on the planet, but it all takes time.**

as the glacier moves past the valley wall and the ice penetrates the rock like fingers. As the glacier moves on, the rock is torn away, trapped within the ice. The Franz Josef Glacier can carry huge boulders intact for several kilometres. High above the valley wall freezing conditions may shatter rock, which then tumbles to the edge of the glacier and adds to the dirty streak along the edges.

The beauty of the ice hides a destructive force that sculpts the floor of the glacier, shaping the very valley in which it sits. As the ice advances, sharp angular rocks embedded in it act as rasping teeth, tearing apart the bedrock beneath. When the glacier has melted, scratches in the rock clearly show the direction in which the ice was going.

Under the glacier, the pressure is so great that it turns the ice to water even though the temperature is way below freezing. The whole of the Franz Josef Glacier is gliding on a film of water that both supports and lubricates its movement. Warmer temperatures melt more ice, making the glacier move even more quickly. This explains how a glacier that is frozen and embedded into the edge of the valley wall is able to move, tearing the attached rock away. It also explains why many glaciers, like the Franz Josef, move faster in summer than in winter.

Under the weight of the ice, the combination of water and sediment is very erosive, carving potholes and channels in the bedrock. On the surface of the Franz Josef Glacier about two-thirds of the way down the ice

sheet, water gushes out of holes with great force, forming streams that flow for a while before disappearing again beneath the ice. Sudden explosions and outbursts of sub-glacial meltwater can break through the surface with little warning and quickly remove huge volumes of sediment from beneath the glacier. Debris carried by the water is left scattered over the surface.

At the end of the glacier, where it is melting rapidly, are huge growing mounds of rubble that the glacier has carried down the valley. A river gushes out from beneath. Its colour is a milky white, rich in finely ground rock and sediment – the result of the ice's powerful erosion. Relics of previous mounds of glacial debris are visible several kilometres further down, where the glacier front settled in the cooler past. Here the advancing glacier has literally bulldozed the shattered rock remains forward. Seen from the air, it looks like a long, thin, horseshoe-shaped hill or mound, green and lush with trees and vegetation growing in its rich deposits.

So how fast is it moving? On 29 October 1943 a Fox Moth, ZK-AEK, a light plane, crashed into the Franz Josef Glacier. Fortunately the pilot and its passengers survived unhurt, but although the plane was recovered, some of its parts, including a cowling, were missing. They finally reappeared, unmarked, at the glacier front in May 1950, nearly seven years later. This episode demonstrated that the ice was moving about 1.5 metres (5 feet) a day. This figure is very close to the average speed for the Franz

Retreating glaciers

All glaciers respond to climatic change. As the temperature increases, even if by only a few degrees, a glacier will retreat significantly within just a few decades. In the mid-nineteenth century, the Franz Josef Glacier covered the lower part of the valley.

140 years ago

75 years ago

Josef, but velocities between 2.3 metres (7 feet) and 7.9 metres (23 feet) a day have been recorded, illustrating how it varies over the seasons and years. Some glaciers are known to move 100 metres (330 feet) in one day.

So the position of this or any other glacier is far from fixed. Over no more than the last century or so, we can see from our Time Machine that the Franz Josef Glacier has fluctuated up to 3 kilometres (2 miles) up and down the valley in response to varying temperatures and levels of snowfall. When the snowfall in the upper glacier is equal to the amount melted lower down, the glacial terminus remains in the same position. An increase in snowfall or a decrease in melting, or both at the same time, will tip the balance in favour of the glacier advance. Of course, the converse is true if there is a reduction in snowfall or an increase in melting. However, the advance is not instant – it takes about five years for the Franz Josef Glacier to start responding to any changes in the climate.

From lush forest to desert plain

Explorer Laszlo Almasy, made famous in the film *The English Patient*, made extraordinary discoveries in a cave in Tassili, Algeria. He found pictures drawn on the cave's walls that clearly showed people swimming. But when he stepped outside the cave he found himself in the Sahara – the biggest desert in the world, with no water around for many kilometres. Later explorers found paintings elsewhere in the desert that showed hippos and crocodiles, water-dependent animals that could not possibly survive there now. So what has happened here?

Travelling back in time just a few thousand years, we find that this area was not a desert but a well-watered place where crops and people thrived. Scientists now think that it turned to desert quite quickly, in as little as a hundred years.

When the Sahara was lush with life, it got its rainfall in the same way as the modern-day monsoon rains in India, which is on approximately the same latitude. In summer, the land absorbed a lot of heat. The hot air rose above the baking land. Moist, warm air above the Atlantic Ocean rushed in to fill the gap, turning to rain when it hit the land and creating some large lakes in the Sahara where crocodiles lived.

But this cycle was dependent on the sun heating the land to a certain degree, and the sun's intensity is not as constant as we may think. The Earth's orbit around the sun varies subtly over tens of thousands of years, sometimes being more circular, at other times oval. These subtle changes become more obvious when you speed up time.

The degree to which the Earth is tilted towards the sun varies due to a wobble in its orbit, and this affects the amount of sunshine hitting any given spot on the Earth's surface. Very subtle changes can have dramatic effects on the climate and landscape.

55 years ago   **40 years ago**   **Today**

When the sea takes over

Many of us love living on the coast, and a large part of our civilization lives within a few metres of sea level. Importing sand to the beaches helps to reinstate artificially California's renowned but fast-disappearing beaches such as Santa Monica, Coronado, Mission Beach, Newport and Huntington Beach. More than 86 per cent of California's 1750-kilometre (1100-mile) coastline is vanishing at a rate of about 30 centimetres (1 foot) per year. It is estimated that as many as 87,000 houses are in danger of crashing into the sea within the next 60 years. Because of erosion, individual houses and developments that were originally built quite a comfortable distance from the ocean are finding themselves perched precariously on the edge.

People living on cliff tops may have a perfect view of the ocean, but if they had the benefit of looking at the world from a Time Machine they would see the process of erosion happening before their very eyes. In some places, like **Na Pali Coast** in Hawaii, it can average about 10 metres (33 feet) in a hundred years. However, in some years, a major hurricane can make 100 metres (330 feet) disappear in just 24 hours. That is big enough for anyone to see!

In Britain, the **Holderness Coast** of Yorkshire is disappearing fast, with around 2 metres (6½ feet) being eroded every year. This is largely due to the bedrock being made of till, sometimes called boulder clay, which was deposited by glaciers over 18,000 years ago. Till is composed of clay, or boulders of intermediate size, or a mixture of both.

Coasts that are particularly susceptible to erosion usually have high cliffs and rugged landscapes. They tend to occur on the leading edge of moving tectonic plates, like the west coasts of North and South America. Glacial action can also aid coastal erosion, as in northern New England and Scandinavia; these coasts are usually dominated by exposed bedrock with steep slopes and a high rise, which adjoins the shore. Although these coasts are fast disappearing, the rate of shoreline retreat can vary due to the resistance of bedrock to erosion. The type of rock is an important factor in the speed at which the coast is being eaten away – the softer it is the quicker it vanishes.

The most prevalent places where the coastline has been eaten away by waves are sea cliffs – steep, vertical expanses of bedrock that range from only a few metres to hundreds of metres above sea level. Their upright nature is the result of continuous wave action at sea level which, over a period of time, cracks open the base of the cliff, causing the rocks above to collapse. Cliffs that extend to the shoreline commonly have a gash cut into them where waves have battered and pounded the bedrock face.

On some of these coastlines, most notably in California and Oregon, a thin, narrow covering of sediment forms a beach along the bottom of the sea cliffs. This sediment is sometimes sand, but it is more often cobbles or boulders. Beaches of this kind usually build up through normal wave action, but are then often removed when the stormy season, when waves are larger, combines with a high spring tide. The presence of even a narrow beach along a rocky coast helps give the cliffs protection against direct wave attack and can slow down the rate of erosion.

Living on the edge – eroding coastline catches out the residents along the Californian coast, where continuous wave action at high tide has eaten away the steep sea cliffs. As the rock face continues to crumble under this onslaught, it will be only a matter of time before these homes are completely destroyed.

Cave paintings of hippos in the **Sahara Desert** reveal a time when there was plenty of water in the area; most of it is now long gone. Giraffes, buffalos and crocodiles were also painted, evidence that water-reliant creatures lived here over **6000 years ago.**

About 6000 years ago, the Earth's tilt towards the sun reached a critical point for the Sahara. The region was receiving less sun. As a consequence, the ground was heating up less, less hot air was rising and the winds from the ocean carrying vital rain were considerably reduced. The balance necessary for sustaining life had been shifted. Some plants had too little energy to grow and began to shrivel, and where the plants started to die, the pale bare earth reflected more of the sun's energy, meaning that the land was less hot than ever and even less hot air was rising to be converted into warm, moist winds.

After this the landscape began to change very rapidly. More plant life withered and died, and the soil dried out and broke up. With no roots or water to hold it in place, it soon began to blow away. Within a very short time, the Sahara had changed from a lush, green land to a desert – and all because of a tiny shift in the tilt of the Earth's axis and a tiny change in the amount of sunshine. This process happened all around the world in the desert belt that runs from North Africa through Arabia and northern India and across the Americas.

Today we can still find remnants of the Sahara's glorious days. Ennedi Lake in Chad, where only three crocodiles of the original population survive, is all that remains of a much bigger lake. Some fossilized wood found in the middle of the Sahara Desert also points to its greener past, and in other places the former courses of rivers can still be seen from the air.

Deserts and lush forest seem to come and go. There appears to be a predictable cycle to the drying up and revitalizing of life back into the land – not only in the Sahara but in deserts all over the world. The Earth's ecliptic orbit around the sun goes in 20,000-year cycles, so if we use our Time Machine to travel forward 14,000 years in time, we can see the Sahara and the other great deserts of the world become wet again and flourish as they once did.

The secret forces of the desert

Other great mysteries of the desert have baffled us for centuries. How the ancient Egyptians transported the massive stones they needed to build the pyramids for hundreds of kilometres 6000 years ago is one such mystery. And what of the pyramids' mysterious neighbour, the Great Sphinx of Giza?

It dates from the reign of King Khafre, the fourth king of the Fourth Dynasty, around 2575–2465 BC. This is known to be a portrait statue of the king, and the sphinx continued as a royal portrait type through most of Egyptian history. But the Great Sphinx is made of just one solid piece of stone. It is far too big to have been transported to its location, so how did it come to be where it is today? Rocks of roughly the Sphinx's shape are found elsewhere in this region and experts believe that they cannot have been formed by water. So what did form them?

In the absence of ice and water, another powerful force shapes the land out here in the desert – the wind. It is responsible for the most famous landmarks of the desert – the sand dunes. But wind combined with sand can also be a powerful destructive force, as the people of the desert know all too well. Vegetation protects the soil, but with that gone there is nothing to hold the soil together, so it turns to sand and dust, which are dry and easily moved. The wind takes the eroded sand and blows it into sand dunes.

The sand grains become very rounded and appear frosted because they have so many collisions with other grains; they are quite different to river sand, which is far more polished. Wind-blown sand can cause more erosion and create more sand in the process. Sand grains do not travel very far in the air, but strong winds make them bounce time and time again in a series of hops. They never rise more than 1 metre (3 feet) above the ground, so it is only the lower portion of an obstacle that gets the full blast of the sand; the upper part suffers only the less damaging impact of the finer grains and dust. This phenomenon has given rise to extraordinary-looking features of the desert called pedestals – rock pillars shaped like mushrooms, which can be several metres tall.

There are two types of sand dunes that form in the desert. Barchan dunes are normally found at the edge of a desert where there is some vegetation and little sand. They are crescent-shaped with their points facing downwind. Between the points there is gravel and barren rock. The wind carries the grains from the windward side up the gentle slope to the peak, where they fall down on the sheltered side. The grains from the back of the windward side of the dune are continually being moved to the front and so the dune slowly moves forwards. The larger they are, the more slowly they move. A small one can advance as much as 15 metres (50 feet) a year. If we speeded up time with the help of our Time Machine and watched their progress, we would see these barchan dunes moving across the desert just like living creatures.

Seif dunes cover larger areas of the desert. They are long ridges of sand separated by strips of rock. Each dune can be 40 metres (130 feet) high and 600 metres (2000 feet) long. The rocks are kept clean and polished by winds that whirl the sand to form circular whisk patterns in the sand.

Fine dust can be lifted as much as several thousand kilometres into the atmosphere and can be blown out of the desert altogether, forming a very fertile, fine-grained yellowish brown soil called loess. Some 10 per cent of the world's land is covered by loess. Massive sandstorms in the Sahara blowing fine dust into the atmosphere and out into the Atlantic Ocean have been recorded on satellites surveying the planet.

A powerful sandstorm can strip the last of the flesh off bones. Sand-blasting is used today in cleaning buildings made of stone. Over time, wind-blown sand can smooth and polish desert rocks. And of course given enough time this can happen on a really big scale.

The Bedouin people fear sandstorms more than any other force of nature. Sand can rip their tents and will get into your clothes, mouth, eyes, and ears, and be extremely painful. In the Iraq war of 2003, as many

Several ancient dried riverbeds in the Sahara Desert are visible from the air, revealing a time in the past when this area had a monsoon climate.

This desert landscape would have been much greener then than it is now, as a result of the lush vegetation that relied on the frequent rainfalls.

helicopters and tanks were damaged by sandstorms as by the battle itself.

Some scientists believe that 25,000 years ago the place where the Sphinx now sits was a plateau in which vertical stripes of soft rock alternated with harder rock. Centuries of violent sandstorms gradually eroded the softer rock. Large areas of soft, poorly consolidated rock and bedrock surfaces became extensively grooved, fluted and pitted. The rock was eroded into alternating ridges and furrows parallel to the prevailing wind direction. Wind-carved landforms like this are known as yardangs, of which there are many beautiful examples in the Egyptian deserts. In their raw state some of them look like the outline of a cat-like creature, called mud cats by the locals. Perhaps the shape of one of these yardangs inspired the ancient Egyptians to take the most impressive yardang they found and carve it into one of their most enduring monuments.

Indeed, if you look carefully at some stones in the desert where the wind comes from varying directions,

you see that they form a pyramid shape. Such pebbles are called ventifacts, and one that has been eroded to three curved facets is called a dreikanter. Ventifacts are generally formed from hard, fine-grained rocks such as obsidian, chert or quartzite. I wonder if they were inspirational objects for the Egyptians and incorporated into their similar-shaped burial chambers? They might have thought such a design was well suited to desert conditions where sandstorms are common.

So we can see that the earth changes on a different time scale to that of humans, and it was only recently that we discovered the enormous amount of time involved. Most of the changes happen over millions of years, but some are short and quick and often violent. When we add up the time taken to create the shape of the Earth as we know it today, it is surprising to note that most of the significant changes happened within the last 50 million years. If we look back at our 24-hour clock (see page 10), that's only in the last 15 minutes before midnight.

RIGHT: **Features of desert life, wind and sand can be powerful in shaping arid landscapes. All over the Sahara Desert there are rock formations shaped by sandstorms. Sometimes called mud cats or yardangs, these may be what the Great Sphinx was originally sculpted from.**

LEFT: **A massive storm in the Sahara Desert whips sand high into the air and blows it out over the Atlantic Ocean, covering the Canary Islands.**

Life resonates to the beat of time. It is a fragile thing and never lasts forever. Every living being has its time on Earth, whether it exists for a few minutes or for thousands of years. When that period of time is over, it dies. All life is bound by time in this way. There are simply no exceptions.

For life to continue to exist on this planet it has to reproduce. Life does not create itself out of nothing; it comes from another living being. There is only one time when this might not have been true, and that is at the time of life's creation 3.5 billion years ago. However, as we will see later, there came a point when new life could no longer be created spontaneously from the primordial sea because these life-producing conditions had changed. Thereafter early life replicated, producing enough copies of itself to survive the violent changing world in which it lived. Some organisms changed their genes through mutation and so evolved into new forms that could live in the changing world around them. It was not until about 600 million years ago that sexual reproduction emerged. This allowed genes to adapt more quickly, as they were better shuffled about, increasing the chances of producing successful forms of life and generally quickening the process of evolution.

Over millions of years, life has turned the Earth from a barren landscape into the thriving blue and green world (left) we know today. Bacteria (right) were the first sustainable life forms on the planet, from which all other life has evolved over the last 3.5 billion years.

3 Life

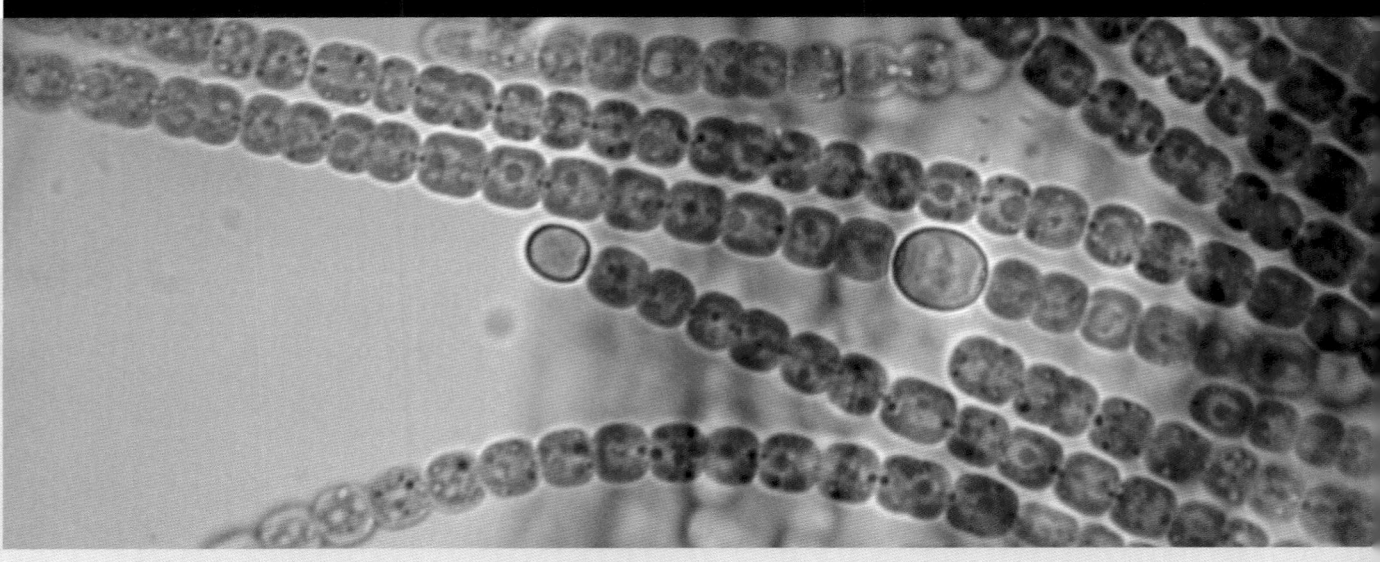

How life forms survive

Life on earth is involved in a constant battle to survive. Some of these struggles we are familiar with, as they happen in our timescale, but others can be revealed only by using the Time Machine. If we slow time down or speed it up, a blow-by-blow account of life's changes through different time warps becomes clearer. As we travel through days, years and centuries we can begin to explore not only how life has changed through evolution, but also how it has changed the planet over time, all in its epic fight for survival.

Time is inescapable for the living. From the moment it is born to the time it dies, every living creature is constrained by the amount of time it has on Earth. In that period, life is influenced by the spin of the Earth to give day and night, by the cycle of the moon that creates the tides, and the cycle of our planet around the sun to give us the yearly seasons. These planetary movements act like a clock to which life sets its rhythm.

If an organism lives long enough there comes a time when its pre-programmed genes will start shutting off vital bodily functions and extinguishing life. Deep within our genetic code there are genes that are triggered to turn off after a certain time, which results in ageing and eventual death. If something has not already killed us, our bodies will.

Life as an entity has been around for a very long time. This is because all new life comes from some pre-existing life – parents – and they too have come from parents. In other words, the genes that programme us to be who we are have existed since life began. Every living creature is the result of some successful reproduction along an ancestral line that goes back a staggering 3.5 billion years. That is an extraordinarily long time for life to be successful and to ensure that the passing on of vital genes from one generation to the next happened without failure. The number of generations involved is incomprehensible to us in mathematical terms. Through this very long time, life evolved to meet the demands of a continuously changing world. What is even more astonishing is that our ancestors, along the way, avoided five mass extinctions when most of life on Earth was killed off.

Only those who successfully reproduce will ensure their species' survival. However, those who cannot breed sometimes help those who can, by contributing to the making of a viable community. For instance, although the workers and soldiers in an ant colony will never reproduce, they are essential to the survival of their society. Their non-reproductive lives are critical in ensuring that those who can breed do so successfully.

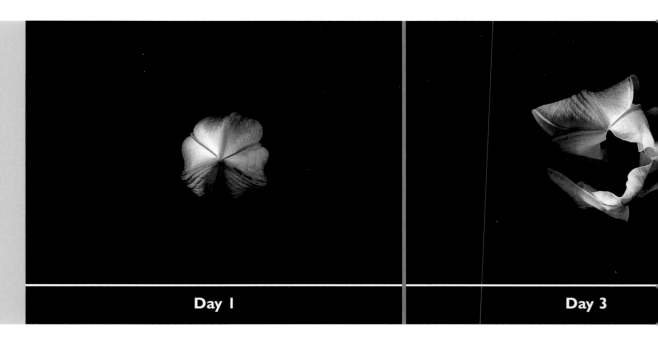

Life's time

All living things are bound by time, like this lily. There is a time to develop, bloom, wither and finally die. In order to ensure the continued survival of their species, all living things must produce enough healthy offspring, which then, in its turn, has to go through an identical cycle of events.

Day 1

Day 3

As we speed through time, it becomes apparent that individuals are less relevant, other than in reproduction, than the survival of the species as a whole. We can see that the most successful creatures are those that adapt to changing circumstances and bring forth new, evolved forms of life. We can also see those forms that completely disappear, that become extinct. Then there are those that hardly change at all, but survive nevertheless. But it is those that evolve that can have a major impact on the planet.

So now we must set our Time Machine on it course. If we can squeeze the history of the planet, all 4.55 billion years of it, into a single day, then we can begin to comprehend the history of life on the Earth.

How life began

It is hard to imagine what Earth was like 3.5 billion years ago. You might at first think that you are on a completely different planet. The land is bleak and barren (very much like the surface of the moon or Mars today), but much more violent. There are far more volcanoes erupting, spewing out sulphur and other poisonous gases. There are continuous earthquakes, and bubbling hot-water pools and mud pools that release steam with a stench of rotten eggs.

There are no blue skies – instead, their colour is dark and yellow. The air is filled with black ash and grit from the continuous eruptions of lava. Bellowing clouds of water vapour are produced from the hot vents that penetrate the Earth. These cool and create rain, which by now has started to form the seas. Thunder and lightning fill the atmosphere with electrical charge. There is very little oxygen in the air, but plenty of hydrogen, carbon monoxide, ammonia and methane. In fact you have to wear a sealed spacesuit with a tank of present-day air inside in order to have any chance of staying alive.

You might think that this is the last place in which you would expect to find any form of life. Indeed, ultraviolet (UV) light penetrates the atmosphere with such intensity that most of modern life could not survive. On top of all that the planet is continually being bombarded by meteors and other cosmic bodies.

But one unique aspect might give you a clue that this is indeed Earth – there is plenty of water. At this time, there are torrential downpours and plenty of lakes and seas. Most of the water percolates through the Earth thanks to volcanic activity. It becomes a rich broth of clay and minerals that are leached out from the Earth, meaning that the oceans contain the trace elements on

Day 5   Day 7

Recreating life

Scientists have tried to recreate similar conditions to those of 3.5 billion years ago in laboratories to see if they could produce a simple form of life. In 1953 Stanley Miller, a young graduate student at the University of Chicago, believed that many of the molecular building blocks of life could be synthesized from a solution of simpler molecules of the sort that probably existed in the pre-biotic seas, by creating similar conditions and passing electrical charge and UV light through the solution.

All Miller's attempts to create new life failed. However, he was successful in creating complex molecules, including sugars, nucleic acids and amino acids – the essential components of life – all in the space of a week.

Since then, other scientists have produced many naturally occurring amino acids and nucleic-acid compounds in the laboratory under pre-biotic conditions. They think that simple pre-biological organic chemicals concentrated in lakes and tidal pools, forming a rich primordial 'soup', could have been the cradle from which

life came. In experiments, the simple molecules, induced by UV light and mild heating, condensed into more complex ones resembling proteins and nucleic acids. If that is possible in a laboratory for such a short period, there is no reason why it could not have happened naturally many billions of years ago.

On top of all that radio astronomers have discovered related molecules in dust clouds far from our solar system. Other similar substances could have arrived from outer space on meteors, adding more complex compounds and all helping to produce basic components for life. This violent, energized state of the planet brought about the construction of the first living material.

The hot springs in Yellowstone National Park, Wyoming, have primitive bacteria growing in them that have been around since the beginning of life on this planet. It is thought that such places were far more common when life began on the Earth, thus ensuring a good start for all early life forms.

which early life will depend. The seas are much warmer than today's, something close to our own body heat of 37°C (98.6°F).

Replication for survival

After millions of years, basic biochemical substances were concentrated together and started to react with each other to form even more complex molecules. For life to continue after becoming established, there had to be long-stranded molecules – polymers – capable of replicating. This vital polymer was DNA, deoxyribonucleic acid, the compound found in every living cell on which life depends for its survival. It is essential for sustaining life for two reasons. It forms the blueprint for manufacturing amino acids for building the living organism, and it also has the ability to replicate itself and so produce more independent individuals of itself. Once DNA could replicate, those molecules best able to cope with their surroundings survived. After certain other basic adaptations were achieved, the first primitive cells arose.

DNA was effectively like a computer program that held instructions for making copies of itself and the cell it resided in. It produced all the parts of the cell, which allowed it to breathe, feed and protect itself, before replicating again. That DNA instruction or code was, of course, the gene. This is the crucial element for sustaining life and it is found in all living organisms, even the simplest, such as bacteria. Even today bacteria remain the simplest and smallest of any living organism, and it is therefore not surprising that they are the oldest fossils ever discovered.

In 1996, Maartin de Wit and Frances Westall made an amazing breakthrough when they were working on one of the oldest rocks known on the planet in the Barberton Greenstone Belt in South Africa. Here they found tiny rice-grain shapes just a few thousandths of a millimetre across. So small were they, you needed an electronic microscope to see them. They were indeed bacteria and they were perfectly fossilized in rocks called cherts, which are formed around intensely active volcanic

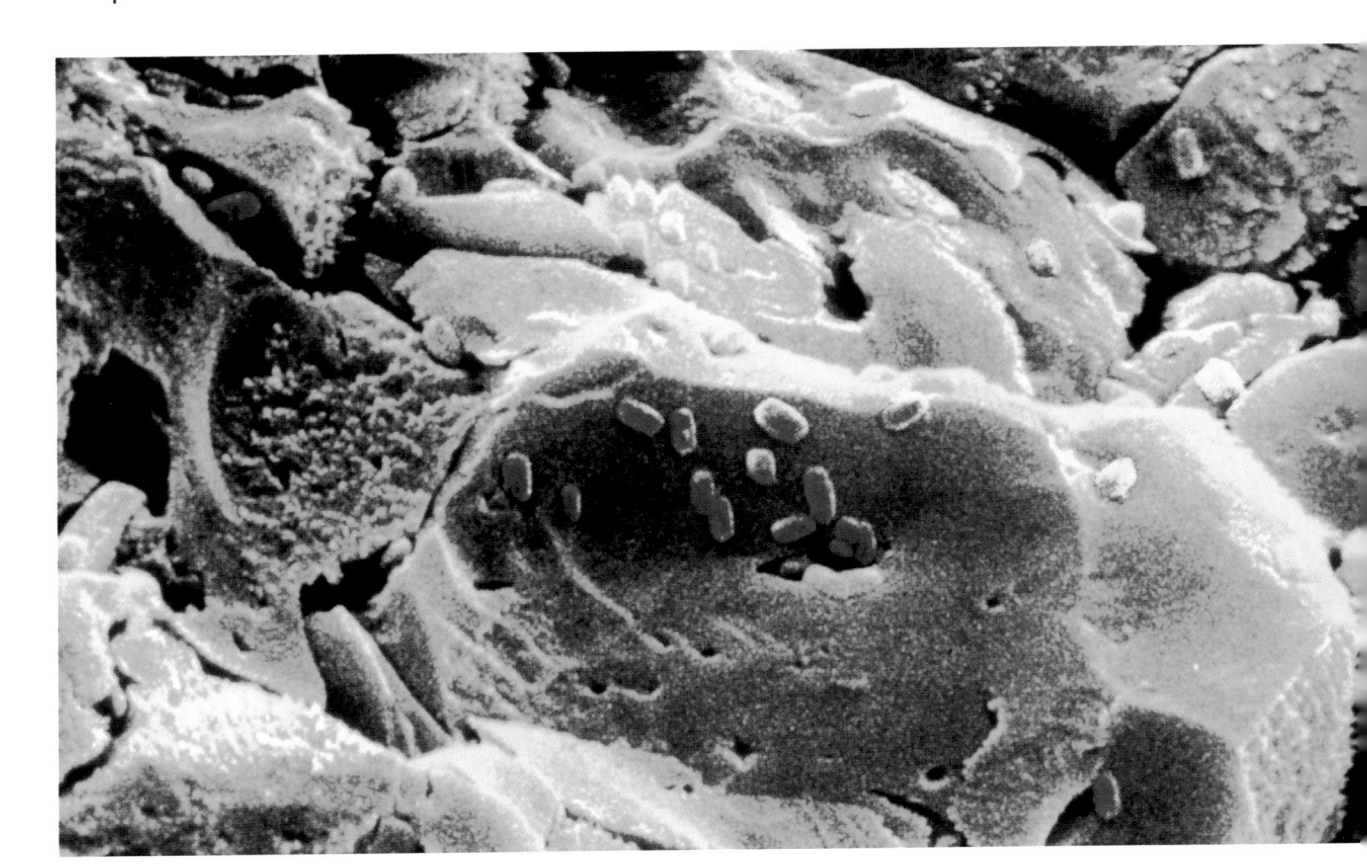

Discovered in one of the oldest rock samples in the world, the Highveld rock system of Barberton, South Africa, these fossilized bacteria, similar to those found in Yellowstone's hot springs today, are estimated to be around 3.5 billion years old.

areas where there are hot springs. These filamentous microbes were dated to 3.465 billion years ago, making them the oldest fossils in the world.

There was a huge expanse of time in which these simple bacteria, known as prokaryote bacteria, could evolve and as time went by they learned to replicate. Some bacteria can produce identical copies of themselves very quickly, in something between 20 minutes and a couple of hours. So their potential evolutionary rate is around 10,000 times quicker than that of humans. Time for a prokaryote is therefore 20 million times faster than it is for a giant redwood, which can live as long as 3000 years. The more quickly an organism can reproduce, the more quickly it can evolve into a new form, which allows it to adjust to changes in harsh environmental and chemical conditions.

With this form of reproduction, evolution happened strictly by the mutation of DNA alone. This meant that any mutation giving the bacterium a better chance of surviving in the future, lived. Any mutations that were not beneficial, and most were not, simply died off. Each generation was so short that evolution was going at warp speed.

The early bacteria lived off energy-rich carbon compounds that were produced over millions of years in these primordial seas. Eventually this food source began to run out and new ways of getting it were necessary for survival. Some adapted by producing their own resources, using the sun's energy in a process called photosynthesis. This required an unlimited source of hydrogen, a gas that is commonly found in volcanic areas.

Simple photosynthesizing bacteria that use hydrogen in this way are still seen in hot springs like those in Yellowstone, Wyoming. They are the most colourful of bacteria or indeed of any life. The variety of colours reveals their diversity. The scalding waters flush up rich minerals from thousands of metres below the surface;

Boulders of stromatolites at Hamelin Pool, in Shark Bay, Western Australia. Stromatolites contain cyanobacteria that were responsible for producing oxygen in the atmosphere 2 billion years ago. This process eventually led to the development of new diverse life forms that needed oxygen in order to survive.

these waters cool at the surface and salts are deposited. It is here that the bacteria are found. They are getting their hydrogen from the sulphuretted hydrogen produced by the volcanic activity below ground. As long as this supply lasts, the bacteria survive; however, it restricts them to living in the one location.

From hydrogen to oxygen

Over time, some bacteria found another widespread source of hydrogen – water. These bacteria were the highest forms of life at the time and are called blue-greens or cyanobacteria because they have a more sophisticated photosynthesizing process using chlorophyll. Today, as a living example of how they must once have thrived, they do well in a very salty inlet called Hamelin Pool in Shark Bay, Western Australia.

Remains of boulders containing living bacteria, known as stromatolites, have been found fossilized in two-billion-year-old rock on the shores of Lake Superior in North America. The ability to extract hydrogen from water meant that these new life forms could, at last, extend their range beyond hot volcanic springs; they were able to flourish anywhere there was a constant supply of water and sunlight.

These new living organisms were to have a resounding effect on the state of the planet. The by-product of photosynthesis using water is oxygen. Early cyanobacteria needed to release the unwanted oxygen that their bodies produced. Oxygen was very harmful to early life forms because it can burn cells from the inside out, grabbing electrons to create free radicals, the highly reactive chemicals that can cause havoc within cells. When oxygen was first produced by cyanobacteria, it literally killed off most other forms of life and became a poison covering the Earth. As a relic of these times, some anaerobic cells – those that do not need oxygen – survive almost unchanged today in swamp muck and in our intestines.

While some of those species that were poisoned by oxygen became extinct, others managed to use this new substance in their metabolism. Oxygen became essential for a new form of life – eukaryotic cells. It was to revolutionize the way most life would survive in the future.

 The time between the emergence of prokaryotic cells and the origin of the eukaryotic cell represents about half of the history of life on Earth, suggesting that it was an evolutionarily difficult transition. If we travel to 2 billion years ago, however, the oxygen content in the atmosphere is starting to increase; and by 1.5 billion years ago, it is nearing present-day levels of 21 per cent. Over the course of some 500 million years, life has completely transformed the gas content of the atmosphere, with monumental consequences.

The production of oxygen gives life new opportunities. Some oxygen molecules produced by these eukaryotic cells react to form ozone (O_3), which is oxygen with a third atom attached, instead of the usual two. The creation of an ozone layer in the upper reaches of the atmosphere considerably reduces the penetration of harmful UV light. This not only turns the skies blue but, more importantly, it creates a less hostile environment that makes the next major development possible – life will soon be able to survive on land.

 About a billion years later, 420 million years ago, marine algae growing around the edge of the sea develop a waxy covering that prevents them from drying out. They eventually free themselves from the sea and go on to become the early mosses and liverworts, forming the first miniature forests. Then begins the slow but steady occupation of land by all other kinds of life. First come the vegetarian millipedes; predatory centipedes, scorpions and spiders soon follow. Life takes a firm hold on the land, stabilizing the soil and giving the Earth a defence from water and wind erosion.

 If we jump forward to 350 million years ago we see the first vertebrates coming ashore. Thus we learn how the planet gradually became blue and green in colour – all thanks to life. Anyone travelling in our Time Machine would at last begin to see some kind of resemblance to our modern-day Earth.

Our planet became green and blue in colour due to new life forms producing oxygen, such as photosynthesizing plants, but it took 3 billion years to achieve it.

Sensing time

All living things respond to our planet's movement, through the cycle of days, months and years. To do so successfully, all creatures must have a sense of time. And the variety of ways in which the passing of time is perceived is quite extraordinary.

How animals perceive time

When we watch the hands going around the clock, time seems to be going at a constant speed. However, our perspective can change according to the situation we are in – time seems to pass more slowly when we are sick or bored, faster when we are healthy and active. Different animals have a different perspective on time, too, and the Time Machine can help us see how the pace of time varies from one species to another. Sometimes, it can be a matter of life and death.

A fly, for example, can perceive much smaller intervals of time than we can. It has very fast eyes, much faster than any vertebrate. This means flies see things happening far more slowly than we do. In Britain, the TV flashes 25 different static images a second; with each image shown twice to smooth out the flashes, this amounts to 50 images per second, which we see as smooth-flowing film. Only if this rate slows down do the images flicker. Flies, however, can see images between 8 and 20 times shorter than that and so would need to see images flash at 400 images per second before they formed a smoothly moving picture. From the fly's perspective, time is moving at a slower rate than ours. This means it sees everything as if it were in slow motion. On the other hand, it doesn't see things that are moving slowly, which gives some of its predators an advantage.

The chameleon is a slow mover, a stealth predator that spends ages positioning itself to catch a fly, with the release of the tongue being a simple reflex mechanism rather than a fast reaction. The fly may not have seen the chameleon stalking up on it, but as soon as the chameleon zaps its long, sticky tongue out, the fly's fast eyes and rapid reaction mean it can take off and be out of the way in a matter of milliseconds. Many more animals that hunt insects have the same disparity in reaction time – frogs and salamanders, for example. They have to persevere and eventually they get their prey by stealth alone.

0.00 seconds

0.01 seconds

Quick reactions

When you see a bird of prey hunting at great speed you may at first think that it sees things slowed down, like the fly. But birds do not in fact perceive time faster than us. Instead they have to process information very quickly, which is a much higher cognitive process in the brain than just taking information in through the eyes. Basically, birds have the anatomy and the physiology to have fast reaction times — we can tell this by watching a sparrowhawk speeding through a dense forest, avoiding branches and twigs as it hunts. Jet-fighter pilots also have the equipment to speed through canyons at break-neck speed, and their reaction times have to be fast if they are to manoeuvre successfully. Some scientists think that if we had wings, we might have reaction times as fast as a sparrowhawk's. But it would all be down to our fast responses and nothing to do with a faster perception of time.

A sparrowhawk is one of the fastest birds of prey. It can speed through forests, avoiding branches and twigs, and attacks at lightning speed.

Taking advantage

The cricket sees time moving at a much slower rate than the chameleon creeping up on it. The reptile then changes its speed, unleashing its long sticky tongue, striking the unsuspecting insect within milliseconds.

Dead Sea shrimps are the opposite of fast-moving flies in the way they see things, because they have very slow eyes. To them, a 0.5-second flash of bioluminescence produced by a protozoan would appear as a slow blur lasting up to what would seem like 5 seconds. No one knows exactly why this is, but one thought is that they see predators better in the dark because the dim light has more time to trigger the light receptors in their eyes.

Reptiles have a problem dealing with variations in time perspective because they have to warm up every morning using the sun's heat before their sight and their hunting abilities are up to speed. In the early winter morning, it looks to the alligators of the Everglades in Florida as if the birds are flying about at a much faster rate, very likely blurred. As the alligators heat up, however, the birds appear to slow down to normal speed and the predators are ready to hunt for anything that moves. Alligators are generally slow movers, but not when it comes to hunting. Their reaction time is faster than any mammal's. Surprisingly for such an ancient animal, their nervous system is designed so that they can snap at prey more quickly than any other vertebrate could react to it.

On the other hand, when male alligators roar during the mating season, they talk to each other at an incredibly slow pace, so much so that it was a long time before anyone realized that they were actually responding to one another. It can take up to 20 minutes for one male to answer another.

Some mammals' reactions are so quick it calls into question whether they perceive the world in a faster time frame than ours. Echolocation requires a bat's ears

Deadly timing

A horseshoe bat catches a moth in complete darkness using echolocation. The bat emits a fast sequence of squeaks and listens for the echo, all within fractions of a second. Such precise timing is required for every successful catch.

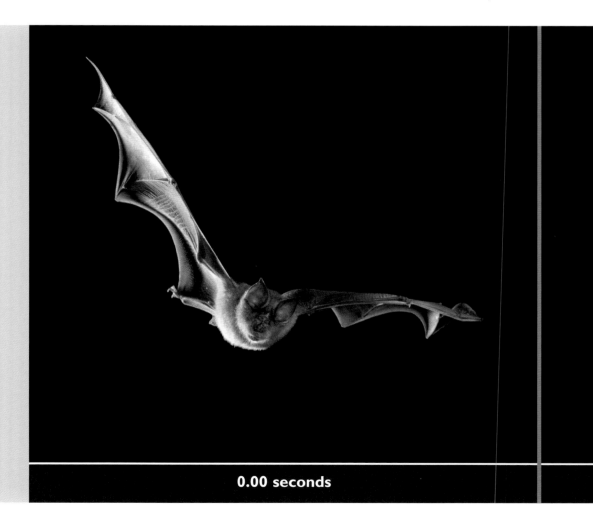

0.00 seconds

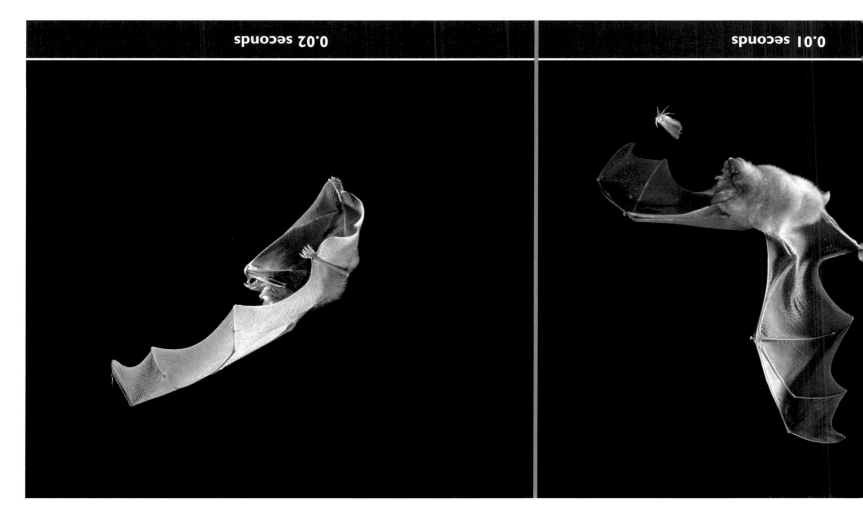

to sense very small time intervals. To map out the immediate environment around them, they listen for the echo of their high-pitched squeaks and work out how far away an object is by timing how long the echo takes to return.

Large brown bats in North America are superb hunters and they use echolocation with time-precision accuracy. Their squeak is so high pitched that they can get an echo from a small object like a moth. As the bat approaches its target, the squeaks turn to higher-pitched clicks, until they are almost continuous. The 'shout' and 'listen' are done in milliseconds. This can cause a problem in that the squeaks can be so deafening that the bat simply cannot hear the echo. Studies done on other bats show that this is overcome by a muscle in the middle ear attached to the trio of tiny bones that transmits the

vibrations of the eardrum to the tubular organ in the skull that converts them into nerve stimuli to the brain. With each click, this muscle pulls the bone aside so that the ear is momentarily disconnected. It is then quickly replaced in time to pick up the echo. Bats' detection of moths is amazingly accurate, and the time intervals in which they process the information are so short that no-one really knows how they do it. However, it is speculated that the brain is able to make calculations and form an audio 'image' almost instantaneously, so it can react quickly to the event. This means that the bat's perspective of time remains the same as ours. Just as we have to concentrate and get our timing right when hitting a ball in soccer or tennis, our perception of time does not slow down; instead our brains act quickly to carry out the correct action swiftly.

Koalas can snooze for up to 18 hours a day. Sleeping reduces their metabolism, thus conserving their energy. This means they can stretch out the time between their meals of eucalyptus leaves.

Slowing down

Many creatures who are active either by day or by night spend their inactive period asleep. Most land vertebrates sleep for at least half the time. Sleep helps to restore energy reserves as well as conserve strength because breathing rates and metabolism are slowed down.

Mammalian predators are among the great sleepers and are active only when they are hungry. Dogs and cats, of both the wild and the domestic variety, can sleep up to 16 hours a day, which means, of course, that they are awake for only one-third of their lives. Many marsupials are even longer sleepers, with the koala snoozing up to 18 hours and the opossum up to 20 hours a day. The larger the animal, the less sleep it needs. Elephants need only four hours sleep a day, giraffes just two hours. So, for these creatures, the day is much longer than that of the smaller opossum.

Some very active warm-blooded animals that depend on food to sustain their energy have found a remarkable way of cheating time when there is very little food about. They use torpidity, which is like suspending time. They go to sleep, but with the added ability to switch off their metabolism and drop their temperature, so that they can conserve their body's food reserves for several hours before going out again searching for a meal.

Hummingbirds are the most energetic of any animal. The rich nectar they take from flowers fuels their hyperactive lifestyle. If this diet is not available to them, they have to be able to switch off their high-burning metabolism, or else they will lose too much energy to survive. They do this by slowing their normal heartbeat of 750 beats a minute down to just 50. Their breathing rate drops too, which suppresses their metabolism. This results in their deep internal body temperature dropping from 38°C (100°F) to below 13°C (55°F). After several hours they have to warm themselves up using the fat reserves they have conserved during torpor, but over all they can save a significant amount of energy during the lean times. They buy extra time by being almost dead to the world.

Many small, high-energy mammals, such as pocket mice and wood mice, do the same when there is little

During its annual hibernation, the common dormouse lowers both its heart rate and body temperature, therefore successfully reducing energy loss. This buys the dormouse more time and enables it to survive long lean periods. Without such techniques, the dormouse would not live through the winter.

food about. This cheating of time might also explain why they live longer. Hummingbirds can live up to 12 to 15 years, about six times longer than any other bird of a similar size.

Hibernation is similar to torpor, but instead of lasting a few hours it is extended to months. Again it is because there is little food around, but rather than just reacting to this, hibernating animals plan it. Their body clock, synchronized to the shortening daylight hours, triggers them to prepare for the big sleep. It was once thought that both hibernation and torpor were primitive ways of saving energy, but they are now known to be a highly advanced physiological way of controlling energy output when faced with any danger of starvation. Bats in particular can reduce their body temperature to just above freezing because they can slow down their metabolism so much that they look as if they have actually died. With the minimum amount of breathing and the smallest ticking over of the heart, they are just able to keep themselves alive, while their own time is held in suspended animation.

Rhythms of time

As dawn breaks, a rose slowly opens, a blackbird joins in the dawn chorus, a hedgehog rustles back into the undergrowth to sleep and a honeybee flits past to visit the newly-opened flower. The rhythm of life can be observed at any time. As we have already seen, we are not the only ones to set our activities to the cycles of the planets – most of life does too. The rhythm of life is in harmony with the beat of time.

The most remarkable image I have seen for a long time was taken from outer space. It was a moving time-lapse image of the Earth, taken over a period of a year, looking down on Africa and Europe. In it, I saw forests, bush and grassland burst into life as the image followed the seasons from southern Africa through to northern Europe. The greening swept north as spring became summer in the northern hemisphere, and turned back to brown as winter hit southern Africa. As summer became autumn in the north, I saw new growth sweeping south as spring arrived in the southern hemisphere. As I looked at this image I realized this was as close as I am likely to get to being in our Time Machine looking down at the Earth from outer space.

What that picture showed most clearly was that all the cycles, from the day to the year, are happening on a global scale. And this is true even of events that take place in a smaller time frame, like high and low tides. At first, it might seem that life is responding passively to these cycles, but on closer examination it can be seen that life proactively anticipates the changes about to happen, so that it is ready to respond to them.

Even the simple cycle of night and day is important in inducing rhythms that affect bodily functions. The influence of the 24-hour cycle is so powerful that most animals and some plants have developed an internal body clock that lines up with this rhythm. This repeated pattern of daily activity is called a circadian rhythm. It is basically internal, built into the central nervous system, and affects sleeping and waking, rest and activity, feeding and general body metabolism, including the daily cycle of hormones that surge at appropriate times of the day to help with those activities.

The Earth's daily light–dark cycle is not necessary for circadian rhythms, but it strongly influences their duration. Even in isolation, away from the daily cycle of light and dark, rhythms maintain the correct time of day. In many cases the circadian sleep–wake rhythm deviates slightly from the Earth's 24-hour cycle. A bird's internal cycle is 23 hours and the human cycle is 25. In all cases the cycle is reset each day by the cues in the environment, such as the sun, so that it is precisely synchronized to the 24-hour day, allowing the circadian system to serve as a true biological clock. Particular internal signals will always occur at the same time relative to the external day. For example, in humans peak levels of the hormone melatonin will be produced in the middle of the night, and the minimum core body temperature will occur in early morning, just prior to waking.

Over the centuries, we humans have come to rely less on our biological clocks, timing our sleeping and waking with the help of the bedside alarm. However, starlings know when it is time to roost just as accurately when guided by their internal body clock. Flying squirrels in North America will arrive at feeding tables at the same time as the night before, within a minute. Regularity is the key to most rhythms.

The internal clock is more than just an alarm clock. Daylight interacts with our bodies and light change stimulates release of hormones. A sense of time is essential for the birds and bees that use the sun as a compass, because the sun's passage across the sky during the day would cause them to fly in a circle if they simply flew at a constant angle to it.

An oak tree seen through the changing seasons is a familiar sight to those in the northern hemisphere. Such seasonal changes are imposed by the movement of our planet and are one of the principle rhythms of time, by which all humans acknowledge and measure the advancement of the year and of time.

Winter

Autumn

Summer

Spring

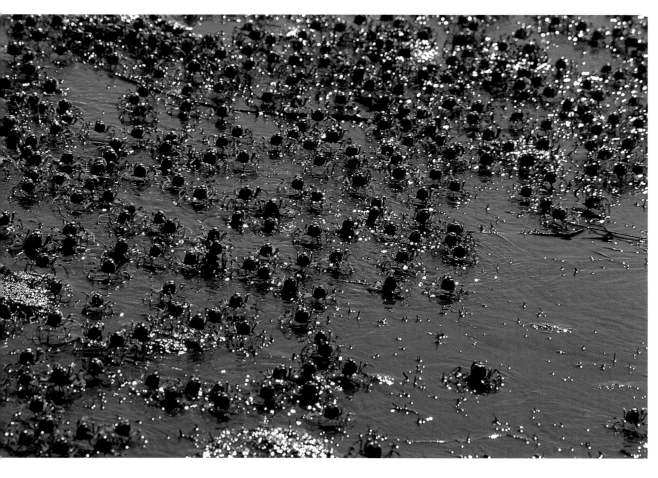

LEFT: **Soldier crabs in Australia storm down the beach to feed at the turn of the low tide. They then rush back to their burrows before the water catches up with them. Their lives are totally dictated by the rhythm of the tides.**

RIGHT: **Thousands of golden-globe medusa jellyfish live in freshwater lakes in Palau, in the Pacific Ocean. They are slaves to the rhythm of the sun. In the early morning they swim up to the surface to absorb the sunlight; in the evening they return *en masse* to the dark depths.**

The internal clock also serves as a way of predicting events. This can best be seen with crabs that have to battle with the rising and falling tides in order to feed safely and efficiently.

For soldier crabs, timing can be a matter of life and death. Off the coast of Australia, tide timetables tell human bathers when to swim in relative safety, but for these crabs swimming is not an option. They cannot swim because, although they have lungs, they cannot hold their breath under water. When the tide comes in they scurry up to the top of the beach and into their burrows, high and dry away from the water's edge. The reason for their warrior name becomes obvious when the tide retreats – all at once a battalion of a thousand or more crabs, stretching up to several kilometres in places, lines up in serried ranks, and in an orderly fashion they charge down the beach towards the sea to feed, some 200 metres (650 feet) away. They have only a short period in which to eat and they cannot waste time, so accurate internal clocks are essential. This type of rhythm is called circatidal.

Soldier crabs feed on detritus in the sand. They do not have brilliant eyesight or any other sensory way of detecting the outgoing or incoming tide. So they must use their internal clocks to anticipate tidal movements. The clock makes them efficient and accurate in their onslaught, and helps them to retreat on time.

Daily cycles

Just as human commuters move in time to the rising and setting of the sun, so too do other animals. In Palau in the middle of the Pacific Ocean, rush hour begins as dawn breaks. The beginning of each new day sees one of the most spectacular commutes in nature – over a million jellyfish rise towards the surface of the salt lakes in which they live. The water resembles a soup of these soft, golden-globed medusae, called *Mastigias*. They have no tentacles with stinging cells and so do not hunt for their food; instead, they are attracted towards the brightening sunlight. The reason for this has only recently been discovered.

Inside each animal are algae called zooxanthellae, that live in a symbiotic relationship with the jellyfish: the host jellyfish gets its food from the algae and the algae get protection from the host jellyfish. The algae get photosynthesize and make food from the sun's rays, while the jellyfish rotate themselves to ensure that all the algae get the full benefit of the sun. At night, the rush hour begins once more as the jellyfish migrate back down to the depths, which hold many nutrients that are vital for their growth and survival.

But these remarkable jellyfish haven't always been daily commuters — they were forced into this lifestyle. Thousands of years ago they freely roamed the sea. As the Palau archipelago grew, the jellyfish were trapped in vast salt lakes; today there are 80 lakes in total. Here, their usual prey of small fish became sparse, so they had to change their diet. They gave up meat and became vegetarian, and changed from hunting to farming. In order to survive, they are locked into a daily rhythm that is in synchronization with the rising and setting of the sun.

Monthly cycles

Just at the right time of night, when the tide is at its highest, the moon is full and if you are lucky there is a cloudless sky, you may catch one of the greatest wildlife spectacles to be seen on any shore. All along the beach for as far as you can see in either direction, hundreds upon hundreds of bizarre marine animals are emerging from the depths of the sea. Flat-helmeted and armoured creatures are coming onshore to carry out their ancient breeding ritual. Although the timing is set by the full

A floral clock

The daily rhythm is not always set directly by the sun; sometimes living things time their activities to coincide with other animals and plants on which they rely for food or reproduction. Flowers are good examples of this. They have regular daily opening and closing times, rather like shops. But they can vary their opening hours to attract the right clientele – pollinators – and beat competition from other flowers. Precision timing is so important to the survival of flowers that they have evolved their own internal clocks to control it.

So accurate are some of these that the flowers can tell us what time it is. At five in the morning, morning glories start to flower to ensure they attract the early-rising specialist bees that feed on them. At 6 a.m. the Californian poppy unlocks its wrapped flowerhead to catch the hoverflies. By nine the dandelions and celandines open, as this is the busy period for honeybees and bumblebees. Noon, and the passion flower unwraps its petals for the hummingbirds and bees. At 4 p.m., the aptly named four o'clock flower opens for the evening. It attracts the hawk moths, which then stay around until 5 p.m., when the Japanese honeysuckle is ready for business. Evening primrose, queen of the night, opens for the bats and long-tongued moths that feed on its rich nectar. If all these flowers were spread around in the shape of a clock face and each flower planted against the right time, we could have a true 'flower clock'.

Again, timing is critical. If neither plant nor pollinator is synchronized to the other, then both can miss out on what they want. Timekeeping like this avoids confusion and saves energy, because those with an appointment do not have to hang around or search aimlessly to see if the other is going to turn up. In this context, timing has become the very essence of living.

Just as their names suggest, the morning glory flower (above left) unwraps its delicate petals at dawn, while the evening primrose (above right) opens for business after dusk. Both flowers are strictly synchronized with the daily habits of the insects and bats that feed on them, who have their own set times of activity. Such precise timing ensures the continuing survival of both the flowers and their pollinators.

moon every spring, you could travel back 500 million years, to a time before there was any life on land, and see the very same incident occurring. These strange-looking creatures have a fixed appointment, which they keep like clockwork every month and which has not altered over millennia.

They are horseshoe crabs, and over three successive nights they struggle out of the sea to lay their eggs on a beach along the Delaware coast of North America. In fact, they are not crabs at all but belong to an ancient group, closely related to scorpions, ticks and land spiders. They use their primeval instincts to plan the safe birth of their offspring, and those instincts are strongly linked to an internal clock.

It is important that the horseshoe crabs coincide their egg-laying with the high spring tide. They can do this only by synchronizing themselves to the patterns of both the sun and the moon, because when these are aligned they produce the spring tides. It is then that the sea comes to the highest place on the shore so that the crabs can spawn, leaving their eggs at the high-water mark, safe from predatory fish.

Looking like miniature army tanks, they clamber up onto the wet sand with slow jerks of their glistening shells, or patrol the shallow water where waves are breaking gently. There are so many of them that they have to climb over one another or collect in small clusters. Some cling to the shells of others and are carried along by the lead animal. The females bury themselves in the sand and lay their eggs, before turning around and returning to the sea. By the time of the next spring tide, a host of fully developed miniature horseshoe crabs will hatch and be washed out to sea.

For a few days each month, at the same time, the

Horseshoe crabs coming ashore at Delaware Bay, New Jersey. They lay their eggs once a year, at a time synchronized with the full moon and spring tide. This ensures that their eggs are left high and dry above the tide line, away from sea predators.

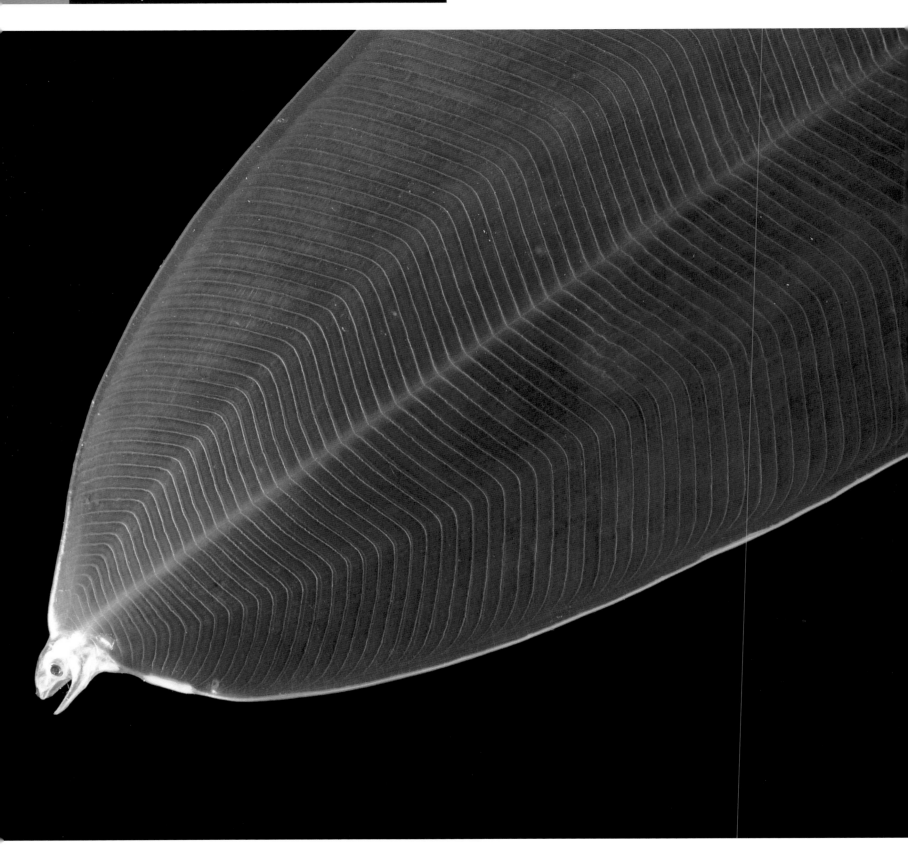

coastline becomes an abundant source of protein, as millions of tiny, pearly-green eggs develop just below the sand. Although safe from fish, through the years they have provided many animals with easy pickings, including hatchling pterosaurs and other dinosaurs! But still the breeding ritual continues, an unstoppable primeval urge influenced by the full moon and the sun entering the spring equinox. Today there are lots of predators, but of a different sort – shore birds. Waders and gulls pick off the eggs, but the ones buried deep down survive. It is a case of numbers; the quantity ensuring that, although the predators may take many of them, there are plenty left over that will make it.

Annual migration

One of the most spectacular migrations is that of the European eel. Its spawning grounds are in the Sargasso Sea area of the Atlantic Ocean, to the east of the Bahamas. Once the larvae have hatched, they make an epic journey, carried on the Gulf Stream and North Atlantic Drift, to the coast of Europe, a journey that takes them over a year.

Why do they make such a long and hazardous migration, when they are at such a tiny, vulnerable stage of their lifecycle? The secret lies in their past, and we will travel back in the Time Machine 100 million years to reveal it.

 Back when these eels first evolved all those millions of years ago, we find that the relative positions of the continents are different to those of the present day, and there is only a narrow channel between North America and Europe. Continental drift is just beginning and there is no North Atlantic Ocean. The spawning grounds of what will become the Sargasso Sea are close to the European coastline. But over the last 100 million years, as

Why do European eel larvae migrate so far? Over the last 100 million years, Europe and North America have separated and moved further apart, so the larvae must embark on an epic journey from their spawning grounds in the Sargasso Sea to the European rivers, that are now over 6500 kilometres (4000 miles) away.

Europe and North America have drifted away from each other just a few centimetres (an inch or two) each year, the spawning grounds have become further and further from Europe. To retain their spawning habitat in the Tropics these tiny larvae now have to undertake a 6500-kilometre (4000-mile) journey.

But that's only the beginning. When they reach the European coastline, from North Africa up to Scandinavia, the larval eels change dramatically into the elver stage, before migrating upstream into freshwater rivers. Their transparent leaf-shaped bodies become more slender and eel-like. A hundred million years ago, the young eels would have remained in coastal areas to feed and grow, but as populations boomed and competition for food increased, they were forced inland, to spend their lives in freshwater rivers and lakes. After about 20 years in fresh water, the yellow eels cease to feed and become silver eels. Then they undertake their spawning migration in September and October, on dark moonless nights, all the way back to the Sargasso Sea.

Irregular rhythms

Rhythms are, of course, predictable and for some animals that can be a problem. The rhythm sets a pattern that can be very easy to follow, especially to those who see them as a good meal, like the shore birds that feed on horseshoe crabs.

One animal avoids this problem by remaining hidden away for a very long time before its emergence as an adult. It is one of the longest-living insects and strangely it spends most of that time underground as a larva, waiting 17 years to reach maturity. We can see the way it does this by entering our Time Machine and going back just a few decades.

The year is 1953 and Charleston, West Virginia, is about to be hit by an invasion. It is a warm night all through the eastern parts of North America and a mass emergence is about to happen. The town looks very different from the way it does today. The latest Chevrolet Corvette passes by, driven by a young man with short, oiled hair with a clean, combed quiff. His car, with the latest wraparound windshield, passes a shop window

Cicada larvae spend 17 years buried in the ground, only to emerge in huge numbers for just six weeks, in order to feed, breed and die. It is thought that such an awkward prime number and the long time interval make it difficult for predators to predict when the larvae will emerge, thereby helping the species to survive.

showing the latest fashions. A mannequin is dressed in a flowery afternoon outfit with a swing skirt; it wears white gloves and has a matching white slim handbag tucked under its arm. From an open window above the shop, music blares out the latest song from Bill Haley and his Comets, 'Shake, Rattle and Roll'.

Across the street in a park with trees, something very strange is wriggling out of the ground – not just one, but thousands upon thousands of dark, glistening creatures. On closer examination, they prove to be emerging cicada nymphs pushing themselves from 1-centimetre (½-inch) holes. The milky coffee-coloured nymph with bright red eyes lumbers awkwardly on the ground. His huge claws, once used to burrow, now have quite a different function – gripping the bark as he climbs the nearest tree. Halfway up, he stops and anchors himself in position with his claws.

Then he suddenly goes into a complete contortion as he begins to split his skin and emerge from his outer suit. Underneath he is pale white in colour, with slender forelegs instead of the lobster-shaped ones he had for burrowing and climbing. Although extremely vulnerable in his soft new coat he will change dramatically overnight.

By dawn his shell has hardened and become dark brown. His wings are pumped up and are almost transparent and mica-shiny in appearance. A clear dark W shows up at the wing tips. He flies off into the treetops where you are more likely to hear him than to see him. The sound male cicadas make is deafening – a sort of whirling, burring, buzzing sound. At least four different variations of their sound are known. The females, who have gone through the same process of change, will fly off into the same trees, but they remain silent. The male's sound comes from special corrugated drumheads on either side of the body that are vibrated by powerful muscles.

Both adults live on the sap of the trees and shrubs, using mouthparts that are developed more for piercing and sucking than for chewing. After a week to ten days, the cicadas mate and the female lays her eggs in a tree beneath the soft bark of twigs and stems. She eventually deposits up to 600 eggs before collapsing and dying on the ground, only a few hundred metres from where she emerged just six weeks before.

After the eggs hatch, the larvae fall to the ground, where they bury themselves to hide and brood for the next 17 years. Then in 1970, right on cue, thousands of cicadas emerge from their burrows. In the meantime, life above ground has moved on dramatically – well, at least in human terms.

This time a Roadrunner Superbird roars by, driven by a long-haired man wearing round glasses. The sound of the Beatles' 'Let It Be' blares from his car radio. The shop window has the same ageless mannequin, but she is wearing a short skirt with a huge belt with a chunky buckle, and a tight tie-dyed T-shirt.

By choosing such a long time interval – and an awkward prime number – cicadas make it very difficult for predators to multiply in order to take advantage of these periodic gluts. It's not worth waiting around for 17 years just for one big feast, so it is better to concentrate on more frequent food sources, even if they may be less plentiful. There will still be a few opportunistic predators in the right place at the right time, but with several thousand cicadas emerging at once, these will have little effect on the whole population.

Scientists doubt whether such precision timing could be controlled solely by an internal clock. But buried deep underground, how can the cicadas measure the passing of time so accurately? It is thought that they are actually picking up cues from the roots of plants. By monitoring sap quality, they keep track of environmental changes above ground, and in this way they can 'count' the years passing by.

How long do animals live?

The limits of the life span of each species appear to be determined ultimately by heredity. Locked within the code of the genetic material are instructions that specify the age beyond which a species cannot live, given even the most favourable conditions. However, most living things die before they reach that age mainly because of the influence of the world around them: many environmental factors, such as disease, predation and climate, act to reduce the chances of reaching the upper age limit.

Quantifying the life span of any living thing assumes that an individual's existence has a definite beginning and an end. For organisms that reproduce sexually this is reasonably clear cut: life begins as a fertilized egg with its unique genetic code, and ends when the animal dies, usually as an older mature adult. However, some organisms, such as many bacteria, algae and protozoans, replicate indefinitely by division of a single parent. Here the beginning and end of a life are more difficult to define and so the age of the individual is more blurred. If replication continues indefinitely, a colony forms. So how do you measure the life span of a colony that does not really have individual birth and death, and cannot be aged in the same way?

In animals generally, birth is considered to be the beginning of the life span. The timing of birth, however, differs so much from animal to animal that this is a poor criterion. For instance, in many marine invertebrates the hatchling larva consists of relatively few cells and is not nearly as advanced towards adulthood as a newborn mammal. And even among mammals there are variations. A kangaroo at birth is about 2.5 centimetres (1 inch) long and must develop further in its mother's pouch, hardly comparable to a newborn antelope, which is walking about within minutes. So we have to be careful, when comparing the life spans of different organisms, to account for any variation.

Small mammals generally have shorter lives than big ones. The smaller the animal, the greater is its heart rate. Heart rate can be linked to breathing rate and so to metabolism, which indicates the speed at which an animal lives its life. Mammals, it seems, all have the same number of heartbeats in a lifetime, so a mouse's heart will beat as often in the course of its life as an elephant's, even though one lives only for a couple of years and the other over 50. We are the exception to this rule, as we should, according to our heart rate, die at 25. Perhaps we did in the past, or perhaps our extended life span is due to the size of our brains, as animals with larger brains also tend to live longer.

Reptiles do not follow this rule, as they are dependent on external temperature to induce their activity and so their heart rate, which fluctuates enormously – a crocodile may have a heartbeat as low as 30 a minute when it is cold, and up to 70 a minute when it is active. Reptiles generally live longer as a result. Tortoises are known to live up to 100 years, if not longer, and this is probably down to their lethargic lifestyle.

In the Arctic, some insects live much longer than those in the temperate and tropical parts of the planet. Some primitive insects such as bristletails, springtails and grylloblattids live in the snowfields. Their bodies are usually dark coloured, which can be very helpful because they absorb heat, whereas a light-coloured coat usually reflects it. These creatures all live their lives at a slow rate, largely because their metabolism is on a very slow turnover and their bodies rely on what little warmth they can get from the weak northern sun in order to do anything. A gryllo-blattid's egg will take one year to hatch and the larva up to five years to mature to adulthood. For most other insects elsewhere in the world this whole process can happen within days or weeks. Grylloblattids are flightless because they simply cannot produce enough energy to fly. It is as if time has completely slowed down for them.

The hunter and the hunted

An owl needs to go through dramatic changes in its growth to transform itself from a helpless pink chick to an effective killing machine. While a barn owl takes 12 weeks to transform into an adult, a house mouse takes just four weeks. The mouse is born with little or no hair. The first fur comes through pink skin, the eyes open, the body changes shape and whiskers grow. The newly hatched owl has downy feathers but is also very vulnerable. Usually two or three chicks are born and they rely on both their parents bringing food back to the nest for them. The mother mouse usually has up to nine young and must feed them by suckling. In between this she still has to go out to feed herself.

The young owl has to invest time in developing its wing muscles and hunting skills before it can survive on its own. As the owl is at the top of the food chain,

the number of prey, like these mice, determines its population. Mice have shorter lives because they have to reproduce more offspring in a shorter period of time than an owl does. A barn owl lives for up to four years, whereas the mouse's natural life span is just two years – though it can, of course, be even shorter if it is unlucky enough to be caught by an owl. However, if predation by hunters such as owls stopped, then mouse populations would suddenly increase. This happened with the house mouse in its non-native Australia, where its numbers increased so much that mice became a plague because there was plenty of food in grain stores, and no cats or owls to prey on them. They were reproducing so fast that their original reason for doing so became pointless; instead it created a disaster because they either quickly died from starvation or were killed by farmers.

Quick breeding

Some small animals go into faster warped time. The antechinus of Australia, a small, mouse-like marsupial, has an even quicker life than a mammalian mouse. In fact it is more like a shrew than a mouse: although it eats nectar it is mainly a hunter, which gives it the enormous amount of energy it needs to carry out its hectic life. Both males and females mate and breed only once in their lifetime, so they need to maximize their chances of producing the most, and the fittest, offspring.

The males become sexually mature when they are only ten months old, but they then face quite a challenge – all the females become receptive within a day or two each year and will mate for only a brief period. For the males to be successful, they have to go into a frenzy of sexual activity, trying to lure females down from the trees. They switch from normal activities such as feeding, drinking and sleeping, to having only one thing on their mind. They go into an orgy of sex, going from one female

The life span of the barn owl is twice that of the mouse, because it has to invest time in developing its wing muscles and complicated hunting skills when young.

The mouse, on the other hand, matures early so that it can breed as quickly as possible before it is hunted down by predators, such as barn owls.

to the next in what looks like a race against time. Each night a hard-working male may mate with as many as a dozen females, virtually one after the other, with some mating sessions lasting up to 12 hours.

But there is a heavy price to pay for all this sexual activity. The males become haggard, thin and exhausted. After two weeks, when it's all over, they are so starved and stressed that they die. Within days, the entire male population of antechinus is dead, leaving behind only pregnant females.

But this sacrifice pays off two months later when the young are born. A successful male may have fathered as many as 60 babies, even though none of these will ever know its father. The young attach themselves firmly to the teats in their mother's pouch, like many marsupials. They grow quickly and are soon hanging beneath her body as she moves around foraging, and so keeping up her milk supply. Ten months after birth the young are sexually mature and the whole cycle begins once more. The females die soon after their first motherhood, leaving the next generation to reproduce the following season.

How long do plants live?

Plants, of course, are not aware of time but they still have life spans that vary enormously. Most plants can be classified as annual, biennial or perennial. Annuals grow from seed, mature, flower, produce seeds and finally die all within the space of a year. Biennials live for two growing seasons. During the first, food is accumulated, usually in a thickened root, like that of beets and carrots,

The male antechinus, a small rodent-like animal from Australia, only breeds once in its short lifetime. In this one session it has to race against time to mate with as many females as it can find, before eventually dying of starvation and exhaustion a few weeks later. Two months on a new generation is born and the cycle begins once more.

with flowering occurring in the second season. As with annuals, flowering exhausts a biennial's food reserves and the plant dies after the seeds mature.

Perennial plants have a life span of several to many years. Some are herbaceous, like irises; others are shrubs or trees. Perennials differ from other plants in that their

Eternal life?

Yew trees are one of the oldest plants on the planet and some have been around as long as our civilized world. In theory, given ideal conditions, they could survive forever.

4000 years ago

3000 years ago

food stores are either permanent or are renewed each year, and they may need anything from a year to many years' growth before flowering. The pre-flowering or juvenile period is usually shorter in trees and shrubs with shorter life spans than in those with longer life spans. The long-lived beech tree, for example, passes 30 to 40 years in the juvenile stage, during which time there is rapid growth but no flowering.

Yews are among the oldest living trees in the world, along with the redwood trees of California. The oldest redwood, thought to be an amazing 12,000 years old, is found in Prairie Creek Redwoods State Park. The oldest yew in Britain is found in Aberfeldy in Tayside, Scotland, and is 4200 years old. The reason yews live this long may be found in the way they grow – slowly. Even when young a yew will increase in girth only half as fast as other forest trees because it is laying down hard, close-grained wood. This means it has immense strength in its trunk and branches. It also gives it protection from anything wanting to bore into it, whether it is fungi or insects.

There is another reason for the yew's longevity: it evolves as it grows. Trees grow from meristems, where cells divide and new tissue is formed. There can be up to 100,000 of these in a single tree and each meristem can mutate, making itself different from its parent. Each branch can therefore be genetically different from all the others. This is critical as it means that the branches are evolving as they grow, making it more difficult for fast-evolving insects, fungi and bacteria trying to consume it.

A yew tree can change from being disease-ridden to being healthy by allowing fungal infections to eat up its heartwood, leaving a hollow tree which, such is the tensile strength of the twisted wood, continues to support the heavy crown of leaves. Meanwhile, branches loop down under their own weight until they touch the ground, and there they set root. A young branch may even touch down into a leaf-mould inside the hollow trunk, and then the tree renews itself from within, by growing a new version of itself inside the old trunk. On the other hand, the spread of disease may split the tree into staves, each bowing out to root itself individually, so that a single tree is transformed into a grove of several trees all close by. By renewing itself in such a way yews can live for an extraordinarily long time, which is why they were grown in old church grounds as a symbol of eternity.

Of course, trees know nothing of old age or death: they grow on and on until an accident finishes them off. It is growth itself that makes them vulnerable. Each year a fresh ring is added to the trunk and branches, drawing on the energy provided by the canopy of leaves. Each new ring is larger than the last, while the crown of leaves stays the same, having reached a point beyond which the framework of the tree can support no more. Many mature trees will snap under their own weight, or keel over in storms.

The yew's last trick defeats time itself. The tree simply stops growing. Then there is no increase in girth or no annual ring. Having reached a sufficient size, it remains stable and, barring accidents, it can stay the same size and, in theory at least, could live for ever.

2000 years ago **1000 years ago** **Today**

Change over time

When looking at time over longer periods – more than a thousand years – we see species begin to change, though some do so more than others. Over millions of years, many millipedes, crocodiles and horseshoe crabs have changed very little, largely because they are good all-rounders, able to survive in a changing world. But for most other living things evolution is a must or extinction is inevitable.

The changing shape of horses

There is something eloquent and majestic about a horse galloping free in a field and nothing more exciting than watching it thundering past you at the races. The horse is so well designed to move at any speed, whether a walk, a trot, a canter or a gallop – the fastest pace of all, in which all four feet are off the ground at once. The sheer magnificence of the horse has attracted many people to dedicate their lives and their money to its form. The racing horse we see today is the result of many centuries of fine breeding, carried out by people who were interested in selecting offspring that could run with the greatest speed and stamina. The wild horse looks similar to the racing one, except that it is perhaps not as tall. So how did the horse become one of the fastest mammals to roam the steppes in Europe and Asia, where it is still found today? Its story is a remarkable one and it is all about how the changing climate transformed a dog-sized forest creature into a magnificent runner of the grasslands.

If you travel to Arizona's petrified forest, to a place called Holbrook, you will find huge fossilized trees lying on the ground where they fell many millions of years ago. The tree *Araucarioxylon* does not have annual growth rings; it was therefore a tropical plant. It grew in a swampy equatorial area in the Triassic period, over 200 million years ago. But Arizona is now a long way from the Equator: it has moved north since these trees flourished.

The modern horse has evolved over millions of years, from a timid forest creature into a larger, stronger animal, content to run on the open plains, in order to survive the changing global climate that destroyed its habitat. Evolution operates on a timescale similar to that of geological change.

This is a good place to start our Time Machine, as it takes us on an amazing journey through the evolution of the horse. We need to travel back to just 58 million years ago, to an Arizona swathed in tropical forests. The average global temperature is 14°C (25°F) warmer than that of the present day. Here we see *Hyracotherium* (previously known as *Eohippus*), the first horse, stepping out from behind a tree. It is a dog-sized creature, totally adapted to living in a forest. With four toes on each front foot and three on each hind, it scampers between the trees like a modern small deer, feeding on fruit and soft foliage.

As we speed forwards through time to 32 million years ago, we find the climate has become cooler and drier, transforming the vegetation to something closer to that of the present day. This change in landscape has driven the evolution of *Miohippus*. It has a much longer skull than *Hyracotherium*, and teeth adapted to cope with tougher plant material – it feeds on leaves rather than fruit. But leaves contain lots of cellulose, which is difficult to break down, requiring bacteria and a longer gut. So the ancient horse has had to grow to accommodate the increased gut. It was thought for a long time that horses got bigger in order to escape

The wild ass is a relative of the horse and lives on the open plains of Africa and Asia.

Horse power

As a result of changing habitat, the forest ancestor of the horse developed an elongated neck to eat grass, and long legs to outpace vicious killers who hunted by stealth and speed on the open plains.

58 million years ago

18 million years ago

predators, but this is now considered unlikely. Taller does not mean faster.

23:54 We now travel quickly forwards to 18 million years before the present. The vegetation has changed even more dramatically, from closed forests to open woodland and spreading grasslands. Grasses dominate the world, and the North American plains are formed. We focus on the latest version of our developing horse – *Merychippus*, a fast-running grazer. It is much larger than earlier horses and has high-crowned teeth to cope with its diet of tough grasses. The horse started eating fruit, then moved on to leaves and then grass, as climate changes altered the food available. Its neck grew so that it could stoop down to feed on grass. Fast running is essential to escape the larger cats, like the sabre-tooth lions and cheetahs, and *Merychippus'* legs show a distinct trend of shortening the upper bones and elongating the lower ones. This allows powerful muscles to be concentrated at the top of the leg. A small movement of these muscles causes a large movement of the foot, which moves extremely fast. The evolution of hooves was not a way of getting away from fast-running predators as was once thought. In fact, these ancient horses were ambushed rather than chased by their early predators. The long legs were indeed used for speed, but were primarily

1 million years ago

Today

to help *Merychippus* get from one forest to the next, where it could feed safely. Raising itself on its one enlarged and strengthened middle toe – the hoof – helped to give extra muscular leverage for galloping and so faster speeds. When the horse eventually left the forest it was well equipped to escape with haste from quick predators.

There were many different types of horses evolving in the forests, although at this time they were confined to North America. Diversity peaked around 15 million years ago, during the golden age of the big mammals, when there were no fewer than 13 horse genera.

 We move quickly forward to 4 million years ago and discover that the seasonal, cool climate is very similar to that of the present day. *Equus* – the first true horse – has finally evolved to a zebra-like creature with stripy legs, very similar to the modern horse but with some primitive traits such as stripes and a long mane going all the way down the backbone. During the first major glaciations 2.6 million years ago certain *Equus* species moved into the Old World, crossing the Bering Strait to Asia and eventually on to Europe. About 1 million years ago, *Equus* species were found all over Africa, Asia, Europe and North America, in enormous migrating herds. Equids, the horse family, are now represented by three wild asses, three species of zebra and two true horses – domestic caballus and Przewalski's.

This magnificent creature evolved because physical conditions changed the landscape and the environment in which it lived. However, such changes are not always down to the physical world – sometimes the living world can put pressure on other life to evolve or die.

The hummingbird's story

The sheer magic of a tiny, glistening and colourful bird hovering over a flower is one of the great spectacles of nature. With its wings beating so fast they blur almost to the point of invisibility, just like a large insect, it feeds on the energy-rich nectar deep within the flowerhead. Some birds have specialized in nectar feeding and eat nothing else, except for tiny insects. They are so small that they can sometimes perch on the flower itself.

Zebras, members of the horse family, stay in a tight group to keep an eye out for hidden predators. Their stripes are thought to confuse the animals that hunt them.

The most highly specialized of all nectar-feeding birds are the South American hummingbirds. They have an additional talent that is critical in the skill of nectar collecting – they can beat their wings so swiftly and articulate them so accurately that they can hang in the air in front of a flower without perching on or even touching the plant.

Hummingbirds have long, thin beaks that are shaped to fit only certain flowers; different species have evolved different bills to match the specific flowers on which they feed. Hovering in mid-air to steady their approach and holding their position for a moment, they delicately dip their beaks into the tiny nectarines with fine precision. Their long tongues collect the nectar with quick flicks. Their digestion has become so adapted to their specialized diet that they cannot cope with other foods. The flowers to which they dedicate themselves are such a valuable resource that they have to guard them against rival birds in their territory.

However, evolution has not always been that straightforward. Two hummingbirds from the same species, one female and one male, have differently shaped and sized bills. But why should the two sexes of the same species feed on two different types of flower?

To see what happened we have to use our Time Machine to travel back 10,000 years – a relatively short time both geologically and biologically – to the period when volcanic eruptions were creating new islands in the Caribbean. At this time, one of these, St Lucia, is colonized by plants from South America, and the hummingbirds have followed, with the purple-throated carib being one of the first. Although in other respects the sexes of this species look almost identical, the males are larger by almost a quarter and dominate the females over the flower of their choice. They thus claim access to the more energy-rewarding *Heliconia caribaea*, which

is a bright red colour, leaving the females with the less bountiful red and yellow variety, *H. bihai*. Unlike the flowers of *H. caribaea*, those of *H. bihai* are curved. Over time, the bill of the female becomes longer and curved, so that it can match the shape of the flower and extract nectar from it more efficiently. In the present day, female purple-throated caribs have bills that are a third longer and more strongly curved than the males'. Over time each sex got better at extracting nectar from its own flower.

However, evolution was not just for the birds. In forests where *H. caribaea* is rare, there is a real danger that male hummingbirds may leave the area. If the males go, so do the females, which would not be a good thing for *H. bihai*, as it relies entirely on the female hummingbird for its pollination. So to prevent the males from leaving the area, *Heliconia* has evolved a red-green morph, resembling *H. caribaea*, whose flowers are shorter and straighter, to match the bills of males.

So not only do animals evolve in response to the changing physical world, they also change due to the living world around them. These birds had to adapt their bills to meet the demand placed on them from their food source. Such pressure to change is there all the time and comes from many different directions. In our normal time frame these changes are invisible. Only by using the Time Machine can we see them unfolding. We see species come and go, and we see certain collections of animals and plants having their moment on the planet. It is only when we stand back and view the entire time that life has existed on Earth that we begin to get a new perspective.

The history of life

If you look at life over the whole period of time it has been around, you can see that there have been clear episodes when it took very different but unique forms. Fossil evidence in rocks suggests that there were times when life flourished and others when it dramatically disappeared. What is interesting about these eras is what they tell us about the way life has reshaped itself over time. Biologists have broken these epochs down into ages, with each age being dominated by a group of newly evolved animals.

A hummingbird feeds on a nectar-rich *Heliconia* flower. Females of the purple-throated carib species were forced to develop a longer, more curved bill, to reach into the long trumpeted flower of the *H. bihai*, because the males had dominated the flowers of the *H. caribaea*, which had shorter, more accessible trumpets.

We saw at the beginning of this chapter that the beginning of life led eventually to the emergence of multi-cellular organisms. The period before this happened lasted 3 billion years and is divided into two eras, known as the Achaean and the Proterozoic.

 With the help of our Time Machine we can travel back in time to see the early phases of our planet's history. The Achaean Era begins with the formation of the Earth's crust about 4 billion years ago and extends to about 2.5 billion years ago. The earliest and most primitive forms of life – bacteria and blue-green algae – originate in the course of this era, about 3.5 billion years ago. The Proterozoic Era follows and gives rise to the first creatures to use oxygen and reproduce sexually. This is when living things take on more interesting characters.

 After the Proterozoic comes the Age of Ancient Life, the Palaeozoic Era, from 550 to 250 million years ago, when the first multi-cellular animal appears. One of the phases of the Palaeozoic is the Devonian period, also known as the Age of the Fishes, from 408 to 360 million years ago, in which the first vertebrates come on land: the amphibians, like frogs and toads.

 The next major episode is the Age of the Reptiles, also called the Mesozoic Era, from 190 to 65 million years ago, the last part of which is dominated by the dinosaurs. Their rule produces some of the biggest creatures ever to roam the earth, but after a relatively short period of time their reign ends and they vanish.

 And finally, the Age of the Mammals or the Cenozoic, from 65 million years ago to the present, when the mammals become dominant, although insects and flowering plants are probably more numerous. Each era is separated from the last by a mass extinction when many species died out but from which evolved new and more diverse forms of life.

The shark has survived virtually unchanged since the Age of the Fishes, which ended 360 million years ago.

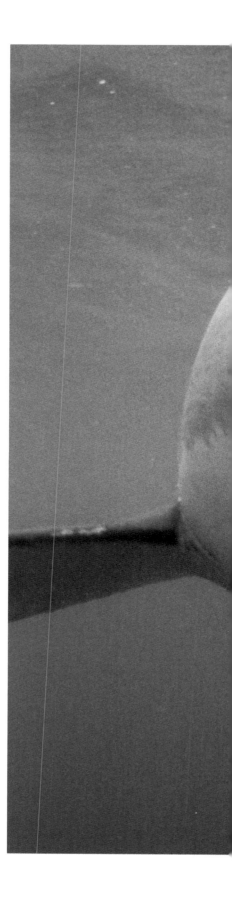

Mass extinction

It is not surprising that disaster can happen quickly and that developments take longer. One of the biggest extinctions ever occurred just before the Age of the Reptiles, between the Permian and Triassic periods, 250 million years ago. Some 90 per cent of all marine creatures and 70 per cent of land vertebrates disappeared. At first examination the extinction seems to have happened in the space of 5 million years. The previous mass extinction occurred nearly half a billion years before, so what had taken over 250 million years to evolve was wiped out in a mere blip of time, geologically speaking – and when new evidence in fossil records was discovered it indicated that the extinction might have taken much less time, only about 165,000 years. So what was the world like prior to the Permian extinction?

Today, the Karoo Desert in South Africa is a stark, mountainous area, with just a few sheep roaming about. But 250 million years ago it would have been a very different place, playing host to creatures we can barely imagine.

 In our Time Machine we can go back to that time, to the end of the Permian period, to see what the Karoo looked like. It is a big, wide river valley. The rivers are gigantic – the size of the Mississippi – lined with plants unlike any we see in the twenty-first century. There are no flowers and no birds. It is a totally foreign-looking world.

Formidable reptilian creatures called synapsids swarm over the conifer forests and fern savannahs in numbers to rival the wildebeest and antelope that roam East Africa in the present day. Alongside them live frogs, turtles, crocodiles and the forerunners of dinosaurs.

In the sea, just 60 kilometres (40 miles) south of the Karoo, life is also flourishing: sharks have already evolved to forms that we see in modern-day waters; there are bony fish, the first animals to have hard bone instead of cartilage; trilobites, giant woodlice-like creatures that live on the bottom of the sea; flower-like animals called crinoids; ammonites; and soft corals.

Then within a couple of hundred thousand years many of these creatures simply disappear. The catastrophe wipes out the very common trilobites and giant sea scorpion, seals the fate of all but two lines of sea urchins and extinguishes most of the ammonites and corals. Nor does the land escape; terrestrial life takes a huge beating too. This is the only mass extinction ever to come close to halting the long, triumphal development of the insects. At the other end of the scale, it destroys the 2.5-metre (8-foot) long reptiles called gorgonopsids and deals a near fatal blow to the synapsids, which had ruled life on land for 80 million years. The pelycosaurs, mammal-like reptiles, some with sails on their backs, are also among the victims. But it opens up the niche for other large creatures – dinosaurs – to take over.

If we travel forward towards the present again, we can climb up a gully in the Karoo to examine clues as to what ensued. As we do, we are essentially climbing through time. The oldest rocks, dating from the middle of the Permian period, are full of fossils of synapsids and other animals, as well as evidence of burrows. But if we go slightly uphill to rocks from the very end of the Permian, we can see that they lack any sign of life. There are no fossils and no burrows. These layers of rock were formed in a total absence of animal life. Some catastrophe must have happened to wipe out almost all the animals at the same time. But what was it?

To answer this we must move to a different continent and look to the time of the Siberian Traps, when an enormous rupture in the Earth's crust caused massive lava flows in Siberia. Here we see a vast wasteland of several hundred thousand square kilometres formed by volcanic lava. The date of these eruptions was 251 million years ago, which coincides exactly with the mass extinction. The volcanic activity here was on a massive scale, the biggest the planet had seen since its creation. Over the course of about a million years, giant vents belched up 11 massive volcanic eruptions, pushing through enough lava to cover the entire surface of the planet to a depth of 20 metres (65 feet). That would have been disastrous for most forms of life.

But the extinction may have been brought about not by lava flows but by the disruption this volcanic activity caused to the climate. Huge clouds of carbon dioxide and sulphates were released, creating a yellow haze that

The nautilus (right) is the sole survivor of a period in the Earth's history over 250 million years ago, when they were very common in all the oceans. Trilobites (above) are now extinct and only survive as fossils.

reflected sunlight, causing the light to dim and chilling the planet. When these gases condensed and fell out of the sky as droplets of sulphuric acid rain, they would have poisoned the ground and destroyed plants and animals.

The Earth's climate changed within a few thousand years from ice age to desert age. As the acid rain and cold clouds faded away, the volcanoes released trillions of tonnes of carbon dioxide, which gradually absorbed heat and created global warming, putting enormous stress on a biosphere that was already badly damaged.

However, it's unlikely that volcanic action alone could have caused the sudden, simultaneous destruction of both terrestrial and oceanic life. A likely additional culprit is shallow water anoxia, which means that the water is starved of oxygen and may explain why almost all marine life disappeared. Another contributing factor was the falling sea level. This was also the time when all the Earth's landmasses had come together to form the supercontinent Pangea, leaving far less coastline, and so less prime habitat for marine creatures to live in.

There is also building evidence of a major impact at the time of the Permian extinction, by either a meteorite or a comet. We already know that a meteorite probably wiped out the dinosaurs in the great extinction at the end of the Mesozoic Era – could this have been a cause of the Permian extinction too? At present there just is not enough evidence for us to be sure, though we are gradually learning more and more as technology improves and it is likely that there was more than one trigger, as one source of stress was superimposed upon another.

Perhaps what is most remarkable about the Permian extinction is how quickly life bounced back again. If we travel to 170 million years ago, a mere 80 million years after hitting rock bottom, we see that living creatures are becoming more diverse than ever before.

Two-thirds of a group of ancient land reptiles, the procolophonoids, managed to escape the devastation, which confirms that it was not as devastating for land animals as it was for marine animals. This probably gave the next land animals a better chance of ruling the Earth.

At the beginning of this new era, the Triassic period, species that can survive in awful conditions have the world to themselves and they spread ferociously for thousands of kilometres. Carpets of bacteria stretch out in the coastal waters, unmolested by grazing animals. A single species of bivalve named *Claraia* races across the shallow oceans of the western USA. On land, patches of quillworts and a few other weedy species replace the diminished lush jungles.

Once life had recovered from the mass extinction, it was changed for good. The ocean's reefs, once composed of algae and sponges, are now hard or stony corals, which still make up the majority of reefs in the present day. Before the extinction, most of the fauna on a typical reef would have been slow-moving animals or ones that were rooted to it – sea lilies, bryzoans and lampshells. Now only a remnant of each group survives. Instead, fish, crustaceans and sea urchins dominate the reefs. The dominant insects had been dragonflies and other species that kept their wings unfolded. But ever since the Permian extinction, insects with folded wings have been more common.

Almost all the synapsids disappeared. Reptiles have become more common, evolving into new forms like crocodiles and turtles. And about 230 million years ago, one slim bipedal reptile gave rise to the dinosaurs, which are now the dominant vertebrates on land. The mass extinction also gave some squat, stubby-legged carnivores called cynodonts the opportunity to evolve. The cynodonts are the ancestors of all modern mammals, including humans. In the future the mammals will have their day, but not until another mass extinction has wiped out the dinosaurs.

Not all extinctions are as catastrophic as the Permian. Over much of the past 600 million years, life on Earth has experienced a steady, low level of extinctions. The average species' life span is about 4 million years, and new species come into existence at roughly the same rate as older species disappear. In fact, more than 99 per cent of all species that ever lived are now extinct. Most of these died out because they were unable to adapt or evolve when change was necessary, and not as the result of a single mass extinction.

How is it that species have a fixed time on this planet? There is not one answer to that question. Perhaps it is because conditions on Earth are never constant. New species evolve, some of which will be better predators, some better at providing food to pass on their sex cells. The physical earth alters, too: new mountain ranges rise up and climate changes. Evolution and extinction are in a continuous and precarious balance. So what about future extinctions – who is likely to survive the next one?

Who will be the winners?

Extinctions will no doubt come about, if history is to repeat itself. The species that have been around the longest may have the best chances of survival, but it all depends on what causes the extinction. If it is anything

The millipede's versatility in the face of global changes makes it a likely contender for surviving well into the future. One of the first animals successfully to colonize land, it has already survived 420 million years on the Earth and escaped two mass extinctions. It has remained virtually unchanged for all that time.

like the last ones, organisms that coped with those catastrophes have a reasonable hope of making it. Living things that are not too specialist in their lifestyle or that can reproduce quickly also stand a good chance. It is difficult to predict the future as we still know so little about the past and present. However, I will stick my neck out and say that locusts, cockroaches and rats are contenders for survival. So is the ancient horseshoe crab, an animal that pre-dates dinosaurs by 100 million years and still thrives today. As we have seen, it has not evolved much in the last 165 million years but it has survived two major extinctions, one before the dinosaurs and one after.

Another likely candidate for the future would be the millipede. Some 420 million years ago, it became the first animal successfully to colonize land. Some modern millipedes are almost identical to their ancient ancestors. Although they are often thought of as having a thousand legs, they actually only have about 200. Each body segment has two pairs of very short legs. Millipedes often curl up into a 'C' shape and remain motionless when touched. They crawl slowly and protect themselves by secreting an unpleasant-smelling cyanide-like compound. There are approximately 10,000 species, most of which live in and eat decaying plant matter; some injure living plants, and a few are predators and scavengers. This versatility could be an excellent attribute for surviving the future.

Then there are the cyanobacteria, the earliest life forms that were responsible for producing the oxygen in the air, making the sky blue and the land green. Inside the stromatolites of Shark Bay in Australia, cyanobacteria are identical to fossil forms that are as much as 3 billion years old. So they are probably one of the oldest survivors on Earth, and they are likely to remain so in the future.

So, despite many global catastrophes and mass extinctions, life has still triumphed over 3.5 billion years. It created our blue and green planet, unique within our visible range of the universe. Throughout the history of life on the Earth many species have made their mark, but today we see one creature making a greater mark than any other, by changing the face of the planet for its own benefit. In the process of all of this it has become obsessed with the master of all change – time. Of course, that creature is us.

Live in any current Western community and the issue of time will always come up, for our modern world is ruled by the clock. We are constantly reminded about time, from the moment the alarm goes off in the morning to the time we go to bed. Everywhere we go there is a clock to remind us of our schedule. Each device we use – the radio, television, mobile phone and computer – tells us the time. The clock is crucial in keeping us synchronized with one another, but there is more to it than that. Our performance in our jobs is measured against time. When we consult a specialist, whether it be a mechanic or a lawyer, the job is priced in anything from hours right down to minutes. By purchasing 'time-saving devices' like vacuum cleaners and washing machines we give ourselves more time to spare. However, in our busy world, even that spare time is soon taken up by some other demand on it.

In the process of constructing our time-conscious world, we have irrevocably altered the face of the planet: turning forests into mono-culture farms, exploiting the Earth's mineral and energy resources and building mega-cities of over 25 million inhabitants. This is unlikely to slow down or stop, with the human population set to double within decades.

LEFT: A field of windmills harvests energy from the wind, to supply us with endless electricity. These man-made structures are symbols of our time that are set to become more common in the twenty-first century.

RIGHT: Combine harvesters reap on a mass scale to feed the ever-increasing human population.

4 Humans

4:1 First invention

Our obsession with time

Rhythm, the beat of time, is a part of living and we humans have rhythm. Start tapping your finger and count the number of taps over a minute or two. It will probably work out to be one tap every half a second. Researchers have concluded that this is the primal pulse within us on which music, dance and even speech are based, and it is consistent with a baby's suckling rate and the speed at which we walk. This internal metronome helps us to stick to any beat once we set it, and so a musical conductor can set a tempo at any rate he wishes, with astonishing accuracy.

According to Carolyn Drake of René Descartes University in Paris, there is a principal beat of 400 milliseconds within us and our brain sets up other oscillations at multiples and fractions of this – 100, 200, 800, 1600 milliseconds. When the second or fourth beat of each bar is emphasized, two pulses are easily excited in the listener's brain, helping us to comprehend the beat, and so the music, better. This beat is important not only for music but also for communication. A group of chimps hooting together has a far greater impact than one shouting alone. This is also true when a number of people sing or chant together. Even clapping has a far greater emphasis when it breaks out in unison. So we perceive the world in pulses, rather than in a continuous stream of consciousness, and this pulse is set by our internal beat. Hearing and seeing rely on contrast because this is the most efficient way to perceive changes, which our metronome samples roughly every two beats a second. Carolyn Drake says we discriminate the world better at this sample rate and it is also the best way of filtering out unnecessary information that would otherwise swamp us. This way we decipher how one event comes after another or the fact that one note in a musical score is longer than another. It is this internal beat that helps us to perceive our constantly changing world and the passing of time itself.

Apart from that, deep inside every cell in our body is another internal clock that helps to ensure we perform to our maximum ability. This clock is set by the daily rhythm of sunrise and sunset, and any variation in that

prepares our bodies for change. For instance many people stack on pounds in the autumn, because their body clock, still in tune with the days when we were dependent on locally available food, responds to the shortening days by telling us to fatten up for the lean winter ahead. The moon sets the menstruation cycle in women and of course the daily rhythm of day and night tells us when to be active and when to rest.

In fact, our body clock keeps every cell in our body on a tight schedule. It dictates when we rest, eat and play and even when our hormones surge. Studies have shown that our bodies respond differently to the external world at different times of the day. We get drunk more quickly when we have a pint at lunchtime and we can do mathematics better in the morning than in the evening. Our internal clocks help us to function at our best.

With or without clocks, we have a perception of time that varies with changing circumstances and as we grow older. When we are young, time seems to go by slowly,

Car lights stream on the streets of New York, famously known as the city that never sleeps. City dwellers live life at a fast pace, where every moment of time is precious to them. To cope with this demand on time, it is increasingly common to find that everything is available at any time of the day or night.

with our lives stretching out before us. As we get older it seems to pass more quickly. One explanation for this is that our metabolism slows down as we grow older. When we are born, our heartbeat is around 140 beats per minute; by the time we are adults it is about 70.

But time perception is based on other factors as well. When we are faced with something new, like visiting a foreign country for the first time, we are taking in more information and it seems that we have done a lot in a short while. It is as if time slows down. This is what many children experience, as they have to take in a lot of information and learn about the ways of the world.

For everyone, it is not just how we perceive time, but

Hormones and chemicals

When we are dealing with something we do not like, such as being ill, time can slow down, too. If we are in a crisis situation such as a car accident, time seems to us to slow down as we begin to react and then respond to the calamity unfolding. Hormones, notably adrenaline, are pumped into the blood supply in order to release energy very quickly. This can make us either fight the situation or take flight and get us out of the spot. We react more quickly, responding to or avoiding things that would normally be too fast for us to see. Distorting our perspective of time in that way can be a real life-saver.

Our perception of time is affected by chemical intake, too. Alcohol usually makes time appear to go past more quickly for the consumer. If too much alcohol is drunk, the drinker thinks he is speaking and behaving normally, but is actually talking more slowly and has very sluggish reaction times, usually with bad co-ordination, so it's not safe for them to drive or operate machinery.

Party time! Drink and drugs can alter our normal perception of the passing of time, which can sometimes be amusing, particularly to observers, but can also be dangerous.

how we use it. Time is an integral part of our lives and today we have more power to manipulate it than we have ever had before. As we saw in the last chapter, every creature has its own life span, a time to be born and a time to die. But in our modern culture, even this can be tinkered with. We can extend our lives by diet, pills and surgery. We can live longer than before, thanks to a better understanding of diet and health. Even when we die, we can extend our existence, by keeping our bodies deep-frozen in the hope that some time in the future we will be able to bring them back to life. We have an obsession with immortality, not only spiritually, but also physically.

Where did this obsession with time start? We saw at the beginning of this book how we went about designing and creating timepieces, with each new generation of clock becoming more and more accurate. But why did we abandon our natural biological and astronomical 'clocks' to take up our own measurement of time? Why did we create the hour, the minute and the second? As we will see, much of this need to measure and keep an eye on time is down to our strategy for survival. To overcome the risk of death, we became more inventive at exploiting the world about us. Every new and better way of doing things theoretically gave us more free time. As communities grew, so did specialization of tasks, so that rather than concentrating on finding or growing our own food, we might choose to become a carpenter or a cleric. As we specialized and invented, we became yet more efficient and so saved more time. And as we developed into better timekeepers, we became more organized and that in turn led to us having a better chance at surviving.

To understand how this all came about we have to take a trip back in time to the very beginning of humankind.

The birth of man

To find out what the world was like when humans first appeared, we need to take our Time Machine back just 5 million years to the great plains of Africa.

For the last few million years the Earth has been cooling. The Antarctic ice has been building up and

Some people are so obsessed with beating time that they want to live for eternity. They request that, on their death, their bodies are deep-frozen in chambers such as this one, in the hope that they may be revived in the future, when advances in science have established immortality as a real possibility.

creating cold-water currents along the west coast of Africa, reducing the amount of moisture in the air and making the climate drier. South Africa is becoming more arid and the forests that rely heavily on rain are gradually vanishing, making way for shrub and grasslands and producing a landscape much like that of the present day.

There are ape-like creatures here, very similar to chimpanzees. Although they are descendants of a forest-living ape that was widespread across Africa, Europe and Asia about 10 million years ago, they now live in the newly-formed plains of Africa. They are known as Southern Apes or *Australopithecus* and they have the beginnings of modern man. Their feet and hands resemble those of their forest ancestors and are good at grasping, an ability that is essential for gathering and picking food. They eat berries and roots, but cannot eat the grass of which there is an abundance. Their limbs are not well suited for running but they are good hunters. They can stand upright for short periods, usually keeping an eye out for predators who could outrun them. If threatened, they use their hands and throw sticks and stones at any advancing animal. Although their brain is the same size as that of their chimpanzee cousins, they have taken an enormous step forward in being able to use stones as weapons.

As we advance quickly in our Time Machine, we see that over the next 3 million years the Southern Ape becomes better and better adapted to living on the plains. With every passing generation his body changes. The feet become flatter, losing their ability to grasp, and slightly arched, making them better at running. The hip joint moves towards the centre of the pelvis and the spine curves slightly, allowing a more upright stance. At the same time the pelvis becomes broader and more bowl-shaped, providing a base for more muscle for both running and keeping upright.

As time goes by, the jaw becomes smaller but the skull doubles in size and becomes more dome-shaped. The whole body is taller, about 1.5 metres (5 feet) in height. This newly-developed creature is known as Upright Man, *Homo erectus*.

It is now 2.5 million years ago. The stones our ancestors threw to defend themselves now play a bigger role in this next stage of humans' development. As tools, they will help speed up laborious tasks and will become crucial to the way humans live their lives and the eventual impact that this is to have on the planet.

Using tools

The need to do things more quickly and easily, combined with our inventive brain, created the first labour-saving devices – stone tools. In order to make these tools, we needed an understanding of our immediate world; we needed to recognize which types of stone would be best for hitting and which would give a sharper cutting edge. Once we knew that, the stone tool empowered us to be efficient and effective hunters. The first known hominid tools date from about 2.6–2.5 million years ago at Gona Kaba, southern Ethiopia. At this time tools were probably quite prevalent, but still crude, since we had not only to learn to make them, but also work out how to use them.

Fascinating studies have been done on how chimpanzees learn to crack nuts using sticks and stones. First you have to get the right stone, next the right anvil, then you place your nut carefully on the anvil and hit it with an accurate blow that has enough force to crack it. The complexity of this task lies in the order in which you carry it out and the various skills that are required at

Just like early man 2.5 million years ago, chimpanzees use basic tools, such as a stone hammer and anvil, in their daily lives. They use these tools to crack **open high-protein nuts with ease and speed. Having saved time in this way, they can socialize with other chimps for longer periods of time.**

each stage. Studies in Guinea, West Africa, show that if a chimp has not attained the skill of hitting a nut with a stone by seven or eight years of age, it never will. Some skills are more difficult, if not impossible, to learn when you are an adult. This must have been similar for early humans, too.

We have to travel in our Time Machine to the heart of Africa, to Olorgesailie in southwestern Kenya, to see the evidence of their highly-skilled handiwork. Here, in a small area, lie some 50 adult ape-like creatures and a dozen young. On close examination we see that these are not humans at all, but members of a giant baboon species that is soon to become extinct. Among the bodies lie hundreds of chipped stones and several thousand rough cobbles. The extraordinary fact about these stones is that they are found nowhere near this place – the nearest are at least 30 kilometres (20 miles) away. The evidence points to Upright Man, who by now is turning stones into

precision tools. Some are carefully shaped with a tapering point at one end and sharp sides that would fit neatly into his hand, like a dagger. The fact that the stones come from so far afield suggests that this group had foreseen the need for such weapons. Baboons are very powerful animals with sharp fangs and the indications here are that these early humans work as a team, using tools to conquer their adversaries. So as early as 2.5 million years ago, our ancestors are already making their mark on the planet and beginning to dominate other species.

They are also using more grunts and calls than chimpanzees, although communication is still crude and mostly done through gestures. The ability to use a wide vocabulary helps early man to speed up the long process of learning, giving us advantages in using tools and coming up with new inventions and discoveries. It is this power to communicate and use tools that leads to the enormous success of Upright Man, whose population is

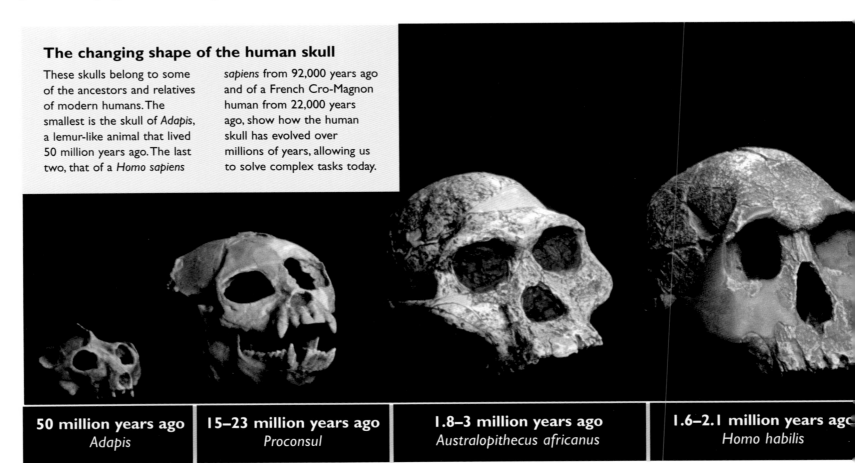

The changing shape of the human skull

These skulls belong to some of the ancestors and relatives of modern humans. The smallest is the skull of *Adapis*, a lemur-like animal that lived 50 million years ago. The last two, that of a *Homo sapiens*
sapiens from 92,000 years ago and of a French Cro-Magnon human from 22,000 years ago, show how the human skull has evolved over millions of years, allowing us to solve complex tasks today.

50 million years ago	**15–23 million years ago**	**1.8–3 million years ago**	**1.6–2.1 million years ago**
Adapis	Proconsul	Australopithecus africanus	Homo habilis

increasing rapidly. As a result, he is soon on the move and expanding out of Africa.

By 600,000 years ago, Upright Man is found across Asia and Europe. The world is going through great cycles of cooling and heating, forcing these early humans to adapt quickly, often using their inventive mind. As we move through time we see more climatic changes, but overall it is getting cooler.

About 110,000 years ago, just ahead of the last ice age, Neanderthal Man appears and is seen alongside early modern humans. Short and chubby, but powerful in build, he has a brain that equals or surpasses that of modern humans in size, even though the braincase is long, low and wide, and flattened at the back. Neanderthals' faces have heavy brow ridges, large teeth and small cheekbones. Their chests are broad and their limbs heavy, with large feet and hands. They appear to have walked in a more uneven, sideways fashion than modern humans. They wear clothing,

hunt small and medium-sized animals like goats and small deer, and scavenge from the kills of large carnivores.

Neanderthal Man is an excellent hunter and tool-maker. His tool kit contains an impressive variety of hand axes, borers, knives and choppers. There are scrapers and heavily serrated blades with a saw-like appearance, essential for working wood, bone and horn into new tools and weapons. However, Neanderthal Man is not to be around long. Why he disappeared remains a mystery – some claim that modern man was more powerful and pushed him to extinction by warring over resources or being better at exploiting them, but we cannot be sure.

By 40,000 years ago Wise Man, *Homo sapiens*, is seen for the first time. He is lighter, slimmer and more athletic, with smaller teeth and a larger cranium to house the expanded brain. The part of the brain that uses speech is fully developed. A skeleton from tens of thousands of years ago is very much the same as ours today.

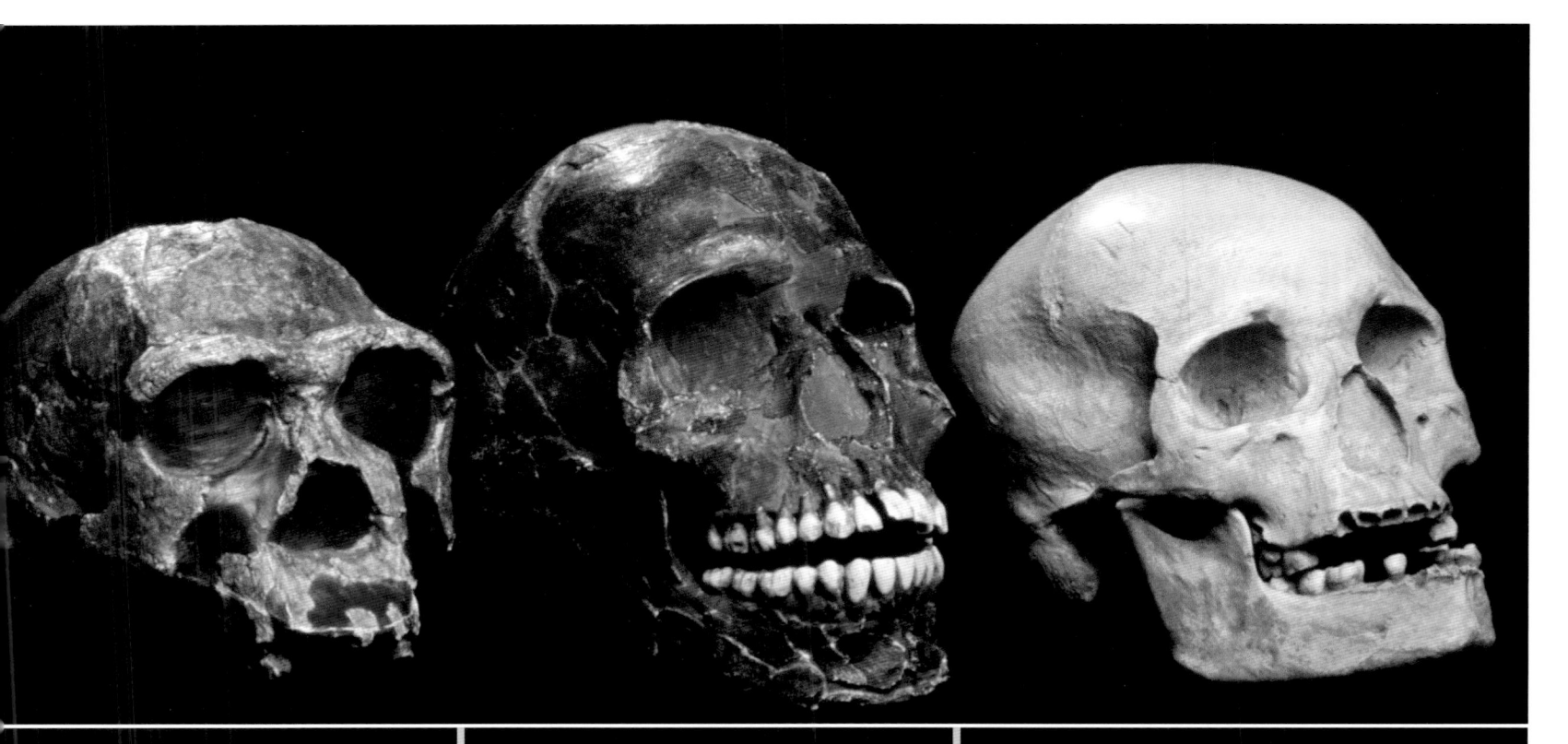

0.3–1.8 million years ago
Homo eructus

92,000 years ago
Homo sapiens sapiens

22,000 years ago
A French Cro-Magnon human

Gradually, the ice-caps are growing bigger, locking up vast quantities of water, and sea levels are falling. With this comes the emergence of landbridges, allowing early man to make his way into the Americas. In Europe he feels the increasing cold, with little or no fur or body hair to keep warm. Using his inventiveness and basic tools, he kills furred animals and strips their coats to wear himself. He also takes refuge in ancient caves. Here, we see the first paintings of the animals he hunted – mammoths, rhinoceros, bison, deer, horses and cattle. Sometimes the paintings are just lines and dots, painted with different colours, a reflection of a vivid and active mind. The artists use tools – sticks and twigs with frayed ends – and local resources, such as the dyes of plants and minerals, to make their work possible. This is the first time that we really see the creative and artistic side of early man's mind at work, and perhaps it also shows us that he has spare time to indulge in these activities.

The progress of invention

But early innovation is slow – it takes over 2 million years for us to progress to a remarkable new type of technology, the composite tool, a combination of several materials. Basically these are tools with handles. This invention greatly improves the power of the sharp stone and makes dismembering larger animals easier. It is also excellent for chopping down trees. Hand-axe blades become progressively thinner, straighter-edged and more effective.

In Schoningen, Germany, the earliest spears can be seen. They are made of wood, but the concept develops quickly to produce formidable weapons. A sharpened stone combined with wood and sinew becomes a spear. The hunter with the spear appears on the scene in Europe more than 30,000 years ago, followed about 17,000 years ago by the atlatl, a hand-held device for throwing spears, which is used throughout Eurasia and the Americas. It is usually a rod or board with a groove on the upper surface and a hook or projection at the rear end to hold the weapon in place until its release. This gives greater speed and force to the projected spear and is used to kill animals as large as the mammoth.

(Modern testing shows that spear throwers can extend the flight from 39 to 69 metres/128 to 226 feet, about as far as an average golfer can hit a distance-designed golf ball with a titanium-shafted metal-headed club.) The combination of atlatl and dart is the human race's first true weapons system, and will be used over a longer period than any other system yet invented. The knack of throwing a projectile at a prey animal with deadly results quickly gives humans an advantage over all other predators. By using these early tools humans are improving their timing with precision accuracy.

The spear and the atlatl are probably the reason for the demise of many wild animals. Tool technology starts to have an impact on the natural world because humans can now hunt larger prey. In Siberia, North America and Australia, huge creatures like the sabre-tooth lion and giant sloth gradually become extinct through hunting and habitat loss.

New developments in toolcraft continue. Some 15,000 years ago, an elongated, multi-purpose flint tool with a cutting edge at one end and a scraper at the other is being manufactured. The burin, as it is called, adds a new dimension to the art of toolmaking. Its prime use is the construction of other tools – a sort of toolmaker's tool that makes it possible to hone fire-hardened antler and bone and to fashion needles from splinters. These sew tight-seamed clothing together with sinew to protect against severe weather. Splintered bone is also formed into harpoons and fishhooks, extending the range of food supply. Burins are also used to carve ivory into decorative figurines.

As time goes by, the bow and arrow replace the spear thrower and become possibly the most enduring mechanical invention of all time. They are a considerable technological advance, being lightweight, powerful and accurate. Triangular arrowheads are also a major breakthrough – a speciality of the Aterian culture at Bir-el-Ater, Tunisia, which spreads along the Mediterranean coast. By 11,000 years ago, a bifacial spear point is being made at Blackwater Draw near Clovis in New Mexico. Even today, the spear and the bow and arrow are essential hunting tools for some people.

The bow and arrow turned us into very effective hunters, allowing us to widen our diet and to use our time more efficiently.

Playing with fire

For a long time it was more convenient to keep a fire alive permanently than to restart it. It was not until about 9000 years ago that we acquired reliable fire-making techniques, in the form of drills and other friction-producing implements, or of a flint struck against pyrites. The simple fire drill is almost universal. Basically it is a pointed stick of hard wood that is held between the palms of the hands, like a sausage between two slices of bread. As the hands move back and forth in opposite directions the stick is rolled or spun. At the same time, it is pressed using the thumb into a small hole on the side of a larger stick of soft wood. The friction between the two sticks creates heat which, when hot, can light dried vegetation. Across the Pacific, Australia and Indonesia, the fire plough and fire saw are variations on the friction method. The ancient Egyptians, Inuit and Asian peoples, and a few American natives all developed mechanical fire drills.

The ability to start fires, rather than just keep existing ones burning, was not only a major technical breakthrough, but also enabled us to alter our environment, to process food and minerals, to provide light and warmth, and to use fire as an energy source.

Another aspect to man's ingenuity was the ability both to make and to control fire. No other animal on the planet has ever been able to do that. This new skill opened another door of opportunity in early man's changing world.

Controlling fires

We have an instinctive fascination with and attraction to fires. There is an awesome quality about them, which we find hard to resist. Their sheer power we both fear and respect. Fire has been a part of human culture for as long as we can remember, but now we know, thanks to archaeological studies, that it has been a part of our lives for much longer.

We do not know how man discovered fire as a tool. But once we found out how to use it, it quickly assumed great importance as a source of heat and light, protecting us from predators and allowing us to cook otherwise inedible food. Fire helped us to clear the landscape of unwanted vegetation and shape it the way we wished it to be – it was fast and there was little effort involved.

The earliest controlled use of fire took place some 1.4 million years ago in Kenya and South Africa. Caves in Swartkrans in South Africa dating back one million years show some of the first evidence of fire use, for warmth and for safety at night from predators. Cooking food by fire allowed these early humans to eat a greater variety of plants and animals, reducing toxicity and parasites. Whether they were able to start a fire is not clear. It seems most likely that they learned how to pick up a burning stick from an existing fire and carry it, then keep it burning as long as they could.

We know that fire was being used in the Zhoukoudian caves in China 500,000 years ago; by 300,000 years ago, its use was widespread around the globe. Of all mankind's early achievements, this really set us apart from any other animal, including our tool-using cousins, the chimpanzees. No other creature on Earth has mastered the skill of using fire for its own benefit.

In due course fire came to be used for more than domestic purposes. Around 40,000 years ago, a herd of over 60 mammoths was ringed by fire and stampeded in a marshy region near Torralba, Spain. Some experts interpret the remaining evidence to mean that the fire was set deliberately and that fire-hardened wooden spears were used to kill the ambushed animals.

Fuel for fires largely came in the form of wood from trees. Forests were felled using the primitive hand axe. If the trees were not replaced, soil erosion led eventually to desertification. Overuse of trees led to an energy crisis 5000 years ago in Khabur Basin in southeastern Turkey. But we soon discovered another renewable fuel – dried animal dung.

Playing with fire can be a dangerous thing and its effects are not always obvious, especially in the short term. When the first people arrived in northern Australia from Southeast Asia 50,000 years ago, they brought with them the knowledge of fire. It was the desert Aborigines' single most important survival tool. They knew how to start it and they knew how to carry it with them. This knowledge was so important that they passed it on from generation to generation, by word of mouth, ritual dancing and song.

The early Aborigines used fire for cooking, light and warmth, but it played a role in almost every other aspect of their life – to produce smoke signals for communicating, to clear the ground of scrub for easy walking and to rid camps and ritual sites of weeds and rubbish. It was used for certain ritual purposes, such as to dispel evil spirits. It served as a torch for anyone travelling at night, to smoke out animals from logs or burrows and as an aid in warfare. It could also strip the ground of plant life to increase the run-off of rainwater and ensure greater water storage in clay pans and small lakes.

Another important use was as an aid to hunting. The Aborigines set fire to areas of spinifex grass so as to direct animals to a spot where they could easily be speared or killed as they fell from a cliff. The day after a fire other animals could be caught because they could be tracked more easily on the cleared earth. Special areas were also burnt with the idea of attracting animals to the new growth that soon followed a fire. This was a sure way of luring animals to a place the hunters knew, rather than spending days tracking them down across the vast landscape.

Clearing the bush

Even today, Australian Aborigines start fires in the bush to flush out game for hunting and to regenerate the area so that new growth will attract more animals. In the past, these continuous activities may have had a profound effect on the environment and the climate. The change was so gradual that generations of Aborigines were unable to perceive the extent of the damage. Yet the whole process of clearing a small piece of land using fire may only take a few hours.

Early morning

Early afternoon

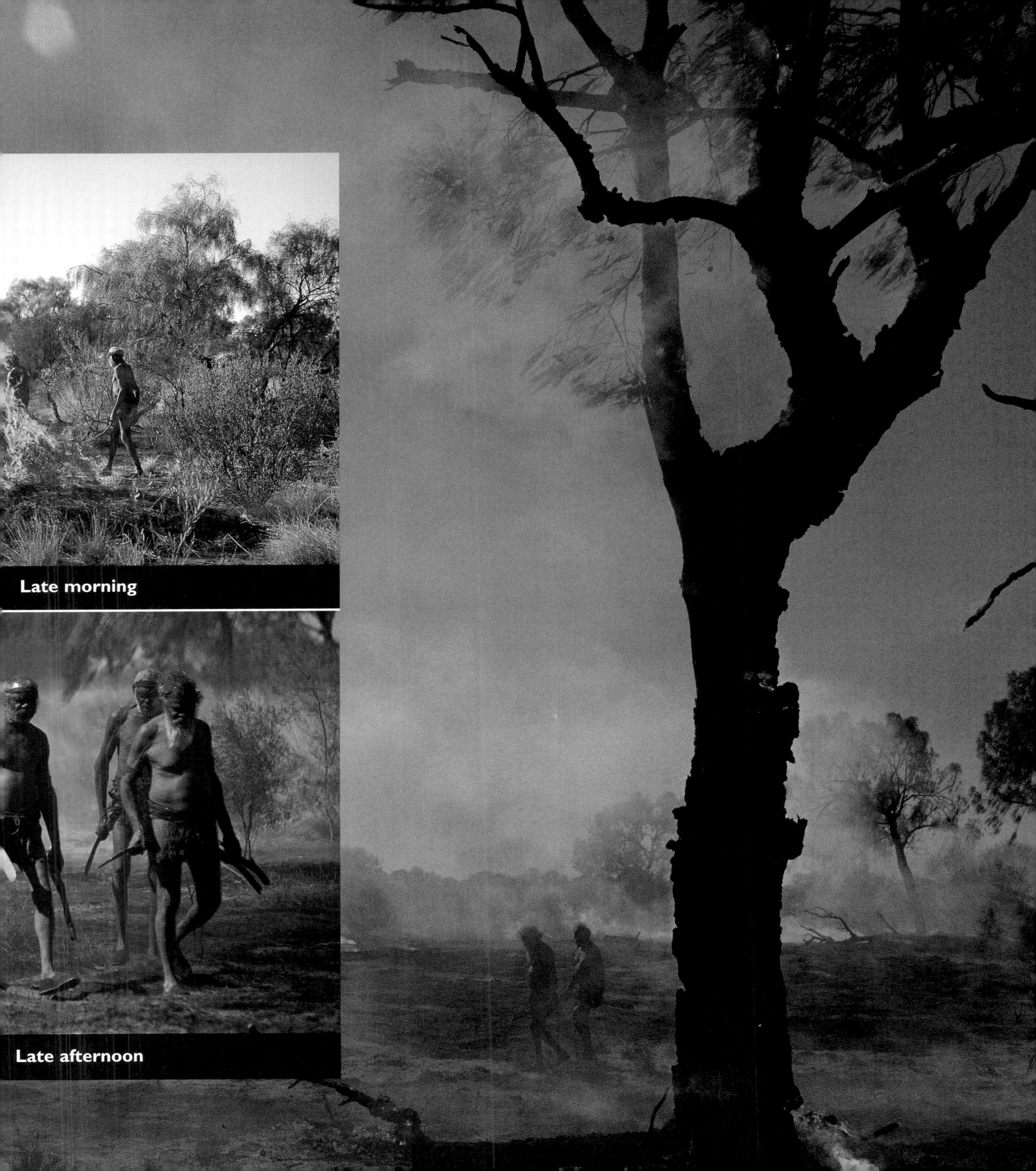

Late morning

Late afternoon

The technique was a great success: it saved huge amounts of hunting time and so was highly efficient. People flourished under this system. Fires were set alight all year round and became far more common than the natural fires started by lightning.

Although it is still a controversial argument, it appears that over time the climate began to change, as the fires themselves tilted the balance of nature. If only these people had had a Time Machine to help them see the damage they were doing in the long term.

The lake that disappeared

At the time, Australia was a rich, green land because monsoons arrived every year, bringing so much rain that rivers formed out of nowhere and ran straight into a huge natural reservoir called Lake Eyre. The vegetation around the lake was abundant and large herbivores were common. Their browsing on the lush floodplains helped produce new growth after the monsoon floods in most years.

But now there were so many fires that the vegetation could not recover quickly enough to sustain the climate that brought monsoons. Nature's delicate balance was upset and the recycling of water between plants and atmosphere was turned on its head. Then 10,000 years ago the Australian monsoons failed altogether. With each passing year the rivers remained dry and the largest lake in Australia gradually dried into a saltpan. All these years later, the monsoons have never returned.

For at least the last 500,000 years Lake Eyre underwent cycles of drying out, interspersed with wet episodes in which the basin was filled by a permanent lake when the monsoon was strong. These were natural cycles related to global changes and were unrelated to human activity. What is different now is that, despite the strong global presence of monsoons, the lake has not filled up.

No one knows exactly how much pressure these early humans put on the environment, but the combination of fires and hunting with weapons must have contributed heavily to Australia losing some 90 per cent of its big animal species within just 10,000 years of the first people arriving. Among the many that disappeared from Australia were giant snakes, giant wombats and giant carnivorous kangaroos.

Australia remains one of the world's driest continents and its present-day vegetation is chiefly composed of flammable plants, dominated by spinifex grass and

Disappearing water

When Australia was visited by annual monsoon rains, Lake Eyre was a real lake. However, as the climate became drier over thousands of years, the lake slowly reduced in size, eventually turning into a dry basin.

10,000 years ago

7500 years ago

eucalypt trees. Fires can spread easily because there are few major obstructions, such as mountains or rivers, to halt them. On top of that, the land does not support large herds of native animals that can reduce the risk of fire by grazing down the grass.

Even today, the relationship between humans, the Australian countryside and fire is so strong that traditional fire ceremonies are still performed. Fire is in the very psyche of the Aboriginal people, used not only for practical purposes but also for their connection with the spiritual world.

Burning the land is a popular technique all around the world. We somehow take it for granted that habitats can tolerate and survive the constant use of fire, but the problem is that we cannot see the scale of what we are doing. Changes that may be very slight in our lifetime accumulate over time, and we can appreciate them only with hindsight.

Just as the control of fire was essential to the development of early man, so it has been essential at every point of the growth of civilization during the succeeding 10,000 years. It was used to turn vessels of clay into pottery; applied to pieces of ore to obtain copper and tin, to combine these and turn them into

bronze around 5000 years ago; and to obtain iron from ore 3000 years ago. Over time, it has been brought under greater and more sophisticated human control.

However, our use of fire on the land has not stopped. In Florida, large parts of the forests around the Everglades are set alight by liquid fuel dropped from helicopters in a supervised operation to prevent wild fires from starting and going out of control, and to help with the process of rejuvenation that would happen if the fires occurred naturally in the first place.

There are still many places in the Tropics where forests are set on fire to clear land for agriculture, much of it on a massive scale. Part of southwestern Borneo has been on fire for a decade due to people trying to deforest part of the jungle. In Brazil in 2002 a patch of rainforest the size of Belgium disappeared in just one year because of large-scale fires. Once the fire has happened it allows the nutrients of the burnt trees and plants to enrich the soil. In the Amazon, however, the deforested soil is so poor of nutrients that local people have continually to move to new areas. The process of clearing the forest and shrub is known as slash and burn, and it is used in preparation for farming, the next major discovery for our early ancestors.

| 5000 years ago | 2500 years ago | Today |

How we took control of our world

The beginning of civilization is not clearly defined, but we assume it began when early man started to settle down and control his environment to grow food and provide semi-permanent shelter. With civilization came man's interest in measuring time, and slowly but surely it became central to everything in our lives.

The advent of agriculture

Some 10,000 years ago, about the time the mammoths were heading towards extinction in Europe, the last major ice age ended and warmer weather returned. Agriculture was slowly taking off as humans started to domesticate crops of wheat and rye, by selecting seeds and sowing them in small plots of land. Both fire and tools were fundamental for farming. Woodlands and forests were cleared using fire, and tools were employed to cut trees, to till the land for planting seeds and to harvest the crop.

Why our ancestors started to farm we will never know. Many scientists believe that hard times pressed the hunter-gatherer into a more sedentary life. Recent studies have begun to show that there was a sudden climate wobble at the end of the last ice age, bringing very unsettled weather. Human populations had increased as the warmer climates produced more food, but any prolonged cold spell could have reduced that and caused famine. Lack of food could have forced our ancestors to use their inventive minds to come up with a solution.

Again no one is really certain how sowing seeds started. Perhaps people who kept old seeds had inadvertently spilled them and later saw them grow out of the ground in spring. To start the process properly they needed to plan ahead, by clearing land, preparing the soil and getting the right seeds for the crop they wanted. Fortunately early humans had the brainpower to do that. This also shows that they had a strong perception of time and particularly what the future was. With this, they were able to predict and plan. They had already seen what they could do to harness nature by using fire and tools. Now they would go further. They would control the production of their own food.

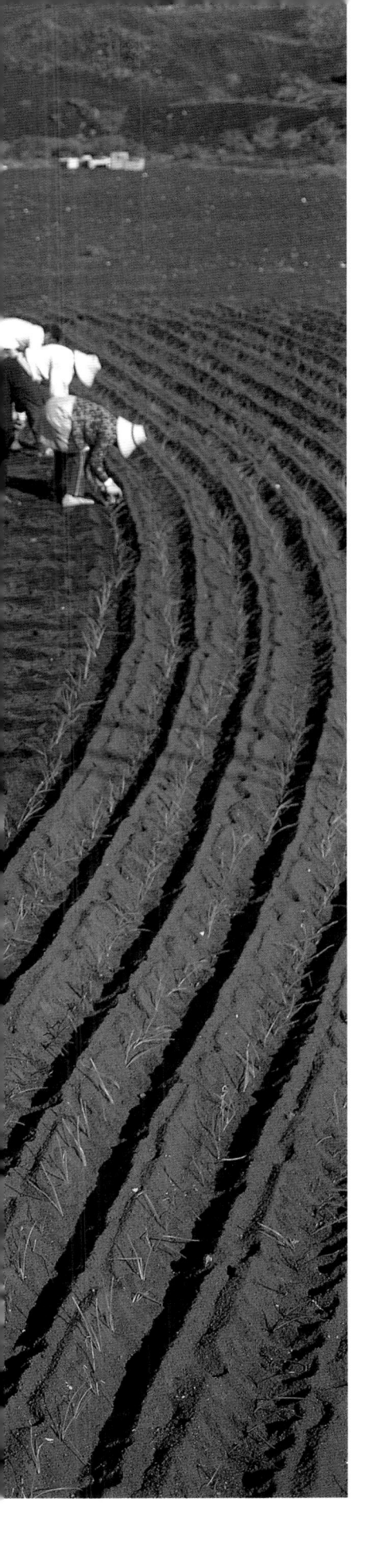

Time-saving farming tools help our ancestors till their crops more quickly, thereby gathering more food than before. When they had too much for their own family requirements, they could sell the surplus, and from these small beginnings the civilized world was born.

Growing your own food rather than travelling many kilometres to find it saves a lot of time in the long run. So the way we used and planned our time began to change dramatically. Instead of putting the hard work into finding ready-to-pick food, we put it into preparation. The benefits were great, as we produced more food with less expenditure of time.

Planning and organizing were now the key to man's survival and were to have an enormous impact on the way we lived. For the first time we had more food than we needed, which brought about a number of changes: we could feed more people, store food for later and even trade it for something we did not have.

We could also afford to settle in one place: we no longer had to be nomads, wandering and wondering where our food was to be found. One of the consequences of this was that we could have larger families. Lots of children were a hindrance when we were moving around, and put both the children and the parents at risk. The overall consequence of larger families was, of course, a growing population. At the time farming took off, the human population around the globe had grown to about 10 million.

Fundamental to being organized and efficient was acquiring particular skills. To be a good farmer you had to know how to do it. Learning how to use tools was vital, as was passing on that knowledge. With the development of our hands to use tools came the expansion of language. Our vocabulary widened as we acquired new skills, probably as a means of communicating important information quickly from one individual to another.

Agriculture started in many places around the globe at about the same time. One of the earliest known sites is in what was once called Mesopotamia, 'the land between two rivers', which lay between the Rivers Tigris and Euphrates in modern Iraq. The rich alluvial soil of the rivers' floodplains provided the right conditions for growing. One of the first crops to be grown, about 9000 years ago, was wheat, but barley and rye soon followed. Farming then spread to the area around the Nile, which flooded the land naturally each year, to provide perfect conditions for producing crops.

Quite separately, rice was first cultivated some 9500 years ago, probably somewhere in India. On the Indonesian island of Bali the way it is grown has changed little from that day. Since rain is not predictable, irrigation systems channel water to flood the fields, helping to ensure a successful crop. Rice-paddy terraces have formed giant stairways into the hills. With the help of simple water-control techniques paddy fields soon covered the land, providing a reliable source of food for the local people.

Since its domestication, rice has been the staple food for over half the world's population and it has fed more people over a longer period of time than any other crop. The land devoted to growing rice covers 150 million hectares (almost 600,000 square miles) worldwide – that's an area almost three times the size of France or Texas. The design of paddy fields has been a hallmark of

agriculture, transforming tropical forests with their multitude of greens into a single-coloured mosaic pattern of flat, watery, terraced fields.

This design was not limited to Southeast Asia. In Machu Picchu in Peru, the Incas shaped similar terraces for maize to preserve the fragile soil. They watered their crop via their own ingenious system of channels, built and carved with stone tools, that fed rainwater down to their maize plants.

Each community around the agricultural world developed a system of carrying water that suited the

Terraced rice paddy fields in Bali, Indonesia, use a simple irrigation system to ensure a bumper crop. Understanding how to alter our environment to maximize growth has been key to our farming success. Such methods have tamed natural habitats, reducing diverse naturally grown crops into a single cultivated one.

landscape, the climate and the crops. Irrigation systems became more complex as populations grew and became more organized.

If farmers could control cereals and crops, there was no reason why they could not control animals, too. Man the hunter had a great knowledge of animals, both predators and prey. Now he was to harness that knowledge into managing them for his benefit.

Taming the wild

The tribes that engaged in hunting and in gathering wild edible plants attempted to domesticate dogs, goats and possibly sheep as early as 11,000 years ago – at least a couple of thousand years before primitive agriculture appeared as a form of social activity. By that time domestication was well under way and over the next 7000 years 14 species of large wild mammals around the world, including cattle, sheep and pigs, were tamed and bred to provide a source of meat, milk, skins and energy. In time camels, oxen, mules and horses were also bred to transport goods and people.

One of the first domesticated animals was the wolf. It is very likely that wolves accompanied hunters and helped them hunt wild animals and they probably also guarded human settlements and warned the inhabitants of possible danger. At this stage, humans probably ate them as well. The wolf was bred for its docile nature towards humans and its skills as a hunter. In the initial stages of domestication we also ate sheep and goats, but later spared their lives when they became valuable sources of milk and wool.

Travelling back in our Time Machine to 7000 years ago, we see the first domesticated sheep around numerous sites of early human habitation in the Middle East, Europe and Central Asia, but these sheep are quite different from the white woolly animal of the twenty-first century. The mouflon, the original wild sheep, is brown with quite short fur, mostly hair with little wool, but humans soon begin selecting sheep with a whiter fleece so that they can colour the wool with their own dyes. Mouflons are widespread across the Middle East and are the ancestors from which all modern breeds of

sheep come. Semi-desert animals that survive on very little water and can tolerate places where there is a high level of salt, they are skittish and will scatter in all directions if threatened; quite the opposite of the modern domestic sheep, which have been bred to congregate together – exactly what you want when you are rounding them up with their old natural predator, the sheepdog.

A natural affinity to hunt sheep was fine when we wanted to kill them, but less good when we started to breed them. So we began to select dogs for their hunting abilities rather than their killing instincts. The skills required for hunting include a herding instinct that allowed the dogs to round up their 'quarry' as well as selecting out individual sheep from the flock. Soay, an ancient breed of sheep found on the island of North Ronaldsay, in the Orkneys in Scotland, feed on the seaweed along the shore. Being desert animals they can withstand the high salt intake. Today they run wild and are not rounded up like most sheep; instead the sheepdogs chase them, pull them down and hold them by the throat. Dogs that kill are put down, ensuring that only those with the right hunting temperament survive. This type of herding probably reflects the early days of domesticating both the dog and the sheep.

Among the many changes brought about by the Roman civilization was the selective breeding of better strains of grain and of domestic animals. The 'Roman' sheep, which had originated in Greek East (Greece and the Middle East) of the Roman Empire, spread throughout Europe, banishing inferior Iron Age species like the Soay to the remote islands of Scotland.

Interestingly, wild sheep grow a combination of wool and hair. Most breeds of domesticated sheep produce wool, though a few produce only hair. Those with fine wool are usually kept for this alone, while breeds with medium or long wool or with only hair are generally raised for meat. Several crossbreeds have been developed that yield both wool and meat of high quality. Breeding helps us get more from one animal, which increases our production of food and clothing. Several hundred breeds of sheep have now been bred in different countries with varying climates and altitudes, according to human demand for clothing and food.

Selecting the right kind of animal for our own use helped us to put less effort into long periods of hard work. This basic act of manipulating nature and the environment fundamentally freed up our time so we could do more with it. Often this led to us producing animals not just for food but for work and leisure.

One animal that fitted the bill was the dog. Early on, we selected different qualities in dogs depending on whether we wanted them for hunting, for protection or for herding. The corgi was originally a cattle herder and its small size was desirable to prevent it being kicked. The Old English Sheepdog was bred to look after and sit with the sheep, and sometimes cattle, as well as to herd them when they were driven along droves for the market from the north of Britain to the south. Other dogs such as the German Shepherd were used to guard the sheep from the wolf. They became excellent guard dogs not only for farmers but also for anybody wanting to protect valuable possessions.

It is only in the last couple of hundred years that we have had sufficient leisure time to play with the way these animals look rather than what they can do as working animals. The poodle was originally a water dog, retrieving game from the marshes where its master hunted. Its name comes from the German *pudeln*, which roughly translates as 'to splash around'. Later, it became one of the first 'fancy dogs', bred purely for their looks. With so much more time on our hands, docile animals became our companions and friends, creatures we could play with, talk to and of course show off to the world.

Domestication has played an enormous part in the development of our material culture, changing both plants and animals to suit our needs. It is precisely those animals and plants that are of most use to us that have undergone the greatest changes from their wild ancestors.

Although the great majority of domesticated animals and plants were selected and developed during the late Stone Age, 9000 years ago, a few notable examples appeared later. The rabbit, for example, was not domesticated until the Middle Ages; the tomato was transformed from an ornamental plant to a consumable fruit from the sixteenth century; the sugar beet came under cultivation for its sugar in the nineteenth; and mint became a garnish as recently as the twentieth. Only in the last hundred years have animals been bred to produce high-quality fur.

The very early production of food and breeding on the farm was to have a profound effect on our lifestyle. As farming became more sophisticated, it began to produce bumper crops. With a surplus of food, trade blossomed – and that was to change where some people lived in the future and the amount of time they had on their hands.

How trade built cities

Early farms developed where the land was fertile and well irrigated. With a surplus of food and goods in these areas people came to trade one product for another, such as a bag of wheat for a goat. Trade routes soon grew and where these paths crossed people traded. Farms were often started here to feed the weary traveller and also because the farmer had a steady supply of buyers for his produce. Different communities bought and sold different things, trading essential resources such as wood, metal, stone and wool, from far and wide.

People settled along trade routes to facilitate access to these valuable goods. Growing markets in specialized goods, along with organized systems of agriculture and irrigation, led to the emergence of the first large towns and cities. These arose in prime locations on trade routes, many on caravan routes or on coasts or rivers. As communities expanded they consumed more resources. Consumption fed back into increased trade and specialization.

Much of this development took place in the Fertile Crescent, a roughly crescent-shaped area of land

Man's agricultural ingenuity changed the landscape and habitat wherever he resided. Thanks to well-managed irrigation techniques, even harsh deserts could become land perfectly suited for healthy crop production. This has led to us being more self-sufficient in extreme environments around the globe.

stretching east from the Nile into Israel, Jordan, Iraq and Iran. In the past it probably had a more moderate climate, better suited to agriculture, than it does today.

One of the first settled communities was at Jericho, near the Dead Sea in what is now the West Bank area of Israel. It claims to be the world's oldest town, but it started off as a village some 9000 years ago. We can trace the development of agriculture to the first early hunter-gatherer communities here. As an oasis, Jericho would originally have been an ideal place for hunting, as the natural spring attracted game. Later, the presence of water would have made it possible to grow crops. Judging by the finely-crafted stone sickles and the mortars and pestles for grinding grain found here, it appears that the town's economy was based on grain production, probably in the fields outside the walls. It is estimated that 2000 people lived in ancient Jericho, making it a metropolis of its time.

It had the added advantage of being situated on an important trade route bringing rare stones such as obsidian, a natural volcanic glass, and shells along the Levantine Corridor, near the eastern end of the Mediterranean. The Mediterranean was a hot spot of trading, because the reasonably calm and protected sea was ideal for the development of shipping and so goods were easily transported from one port to the other.

Around 4000 years ago, the city of Troy in northwestern Turkey lay on a river that gave access to raw materials north of the Black Sea (today it lies 8 kilometres/5 miles from the river). The city controlled the river traffic between the Black Sea and the Mediterranean by allowing ships to dock there for a fee. These ships were often forced to wait at Troy until the wind changed, allowing them to continue their course. The discovery of thousands of spindle whorls shows that the city was a major manufacturing site of wool textiles, where ships came to transport and sell their goods.

Earlier, wool textiles had been crucial in the rise of cities in the Fertile Crescent. The plains had no metal, stone or wood, but produced wonderful textiles at a time when wool was seen as a luxury item. Cities were made up of people who specialized in their craft, whether they were weavers, carpenters or toolmakers. As they became

The city of Troy was once a thriving port, an essential stop-off for all ships entering the Black Sea from the Mediterranean. The exact site of this abandoned city remained unidentified until the nineteenth century, when excavations began.

150 years ago

90 years ago

Boom town

Las Vegas is one of the fastest-growing cities in the world and it is right in the middle of a desert. The reason for its success is simple – the trading of money. Tourism, and particularly gambling, are the mainstay of the economy of this growing city. Up to 30 million people a year visit its bright lights. However, every week a further thousand people come to stay permanently, either to retire or to work, to be part of the dream and its success.

Las Vegas is one of the newest big cities, with plenty of sprawl. Less than 70 years ago it was nothing more than a dusty watering hole in the desert. None of the casinos and themed hotels existed. In 1911 it had only 800 inhabitants and covered just over 4900 hectares (19 square miles). By 1930, it had grown to 5165 inhabitants. With the building of the Hoover Dam only 48 kilometres (30 miles) away, cheap electricity gave it the boost it needed to grow. By the 1980s Las Vegas was booming and the population had reached 368,360. Today it continues to grow at an alarming rate; a sufficient number of houses simply cannot be built fast enough for the new residents. One estimate is that its population will double by 2015.

Las Vegas started off as a small desert town, but with the arrival of cheap electricity from the Hoover Dam, it soon found a new lease of life and grew rapidly. Today they cannot build houses fast enough for the burgeoning population of the city.

more expert they needed more special materials in order to produce their goods, and, thanks to the market economy, these trading places readily supplied their wants.

A city's success depended on what it could offer. Over 2000 years ago, Petra was a thriving city in southwest Jordan. It was built on a terrace by the Valley of Moses, where the Israelite leader Moses is said to have struck a rock and water gushed forth. The valley is surrounded by stunning sandstone cliffs marbled with shades of red and purple varying to pale yellow. The name Petra was given by the nineteenth-century English biblical scholar John William Burgon, who described it as 'a rose-red city half as old as Time'.

Although the original inhabitants are long gone they have left their mark, quite literally – in the huge buildings that were carved into the rock face. The most elaborate of these is called the treasury or Al Khazneh, which is actually a vast tomb. During its construction over 700,000 cubic metres (nearly 25 million cubic feet) of sandstone was excavated to make the central chamber. Ad-Dayr – 'the Monastery' – one of Petra's best-known rock-cut monuments, is in fact an unfinished tomb façade, though it was used as a church during Byzantine times. It is just one of many buildings in what has to have been one of the most spectacular cities of its time. But why was it built in the desert and then abandoned and left to crumble back into the ground?

70 years ago

20 years ago

Today

With our Time Machine we can go back to a period when Petra was thriving, to discover what lay behind its success and ultimate demise. When we arrive we are dazzled by the activity, the like of which has never been seen before. Some of the city's success can be put down to the ingenuity of the Nabataean people who live here. They are able to control a vital resource – water. To support the city's large population, its residents maintain an extensive water system, including dams and rock-carved water channels. Clay pipes are used to bring water to the city, where it is stored in cisterns hollowed out in the sandstone, and then distributed to the houses. In case of sudden flash floods, huge tunnels, nearly 6 metres (20 feet) in diameter, have been dug to divert water away from the city. The Nabataeans even have baths, which make them far ahead of their time.

Under Nabataean rule, Petra prospered as a centre of a spice trade that involved such far-flung realms as China, Egypt, Greece and India, and its population swelled to between 10,000 and 30,000. Traders could not only stock up with food and water here, the city also provided a ready and profitable market for their goods.

Until 1800 years ago, Petra was a boom town, with more and more wealth being lavished on it, but then it was abandoned. It was not war, disease or failure of the rains that led to the fall of this beautiful city; it was the same thing that had made it successful – trade. The Arab caravans that used to pass through, bringing with them their goods and wealth, suddenly stopped using this route. With greater achievements in boat building, ships set sail on longer routes and moved far more quickly, so traders took to the sea. The old Spice Route across the land was reduced considerably as spices were shipped back and forth from Asia by sea and Petra was no longer an attraction. The city fell silent and was lost to humankind until as recently as 1812, when the first Europeans to see it told the world about its secrets and beauty.

From ancient city to modern metropolis

In the meantime, other cities arose with populations that required agriculture to produce food more efficiently and in greater quantities to feed them. City life also led to people being more efficient in their use of time. With this came politicians, managers and clerics who required greater organizing skills. Sundials and water clocks were now being used for measuring the time in cities from Egypt to Greece, calling people to social and political meetings, and to prayer. The cities became a sort of Time Machine where people were able to manipulate their time to suit themselves. Efficiency meant time was saved and spare time could be used however we wanted. Eventually cities would enjoy an active existence virtually 24 hours a day, seven days a week.

Today the buildings may have got bigger and taller but the basic premise of the city is the same. Water and food supply remain high on the agenda, as does specialization in all kinds of skills. It is the combination of various skilled jobs, each one important in running the city, that makes the place work and prosper. Competition in trade puts pressure on everyone to be more efficient, and the more efficient we become the faster we do things. This is why cities are so hectic and why everyone is so time-conscious.

Cities are now growing faster than ever before – it is as if we have revved up our Time Machine and are speeding up time. Yet, as we know, it is the rate of change that has speeded up, not time itself. This growth has produced extraordinary new cities. The most populous in the world is Tokyo with a staggering 26 million people and rising. Mega-cities all over the world are on the rise as more and more people move from the countryside to the cities, especially in developing countries. Why are mega-cities so popular? To answer that we need to go back 200 years or more, to discover that cities were given a major boost by another human revolution, one which was to add greatly to their size. It was also going to change the way we lived our lives and the way we spent our time. This was the Industrial Revolution.

Tokyo is the most populous city in the world, with 26 million inhabitants, and it's still growing. The attraction of living in man-made environments is strong because we can manage our lives there more easily. However, the more time we create for ourselves, the greater the pressure to use it efficiently.

3:3 Industrial Revolution

Man and machines

Our obsession with time has increased rapidly in the last 300 years as our inventive mind came up with solutions that made arduous work easier and carried it out faster and more cheaply. The production of goods on a large scale was a result of the burgeoning age of science, which promised a better world for everyone. Even time itself was seen as a valuable asset.

Mass production

For 7000 years we have tirelessly developed new and increasingly elaborate tools and machines to carry out a vast array of different tasks, each helping us to undertake a particular job more effectively than before and usually in better time. Some of the biggest developments of the early industrialized period concentrated on producing cloth.

Wool was originally made into thread by hand spinning, with the individual fibres being drawn out of a mass of wool held on a stick, twisted together to form a continuous strand and wound on a second stick, or spindle. In the Middle Ages this was replaced by the spinning wheel, a device that speeded up the whole process. The earliest looms for weaving date from 7000 years ago and consist of beams fixed in place to create a frame that holds a number of parallel threads in two sets, alternating with one another. By raising one set of these threads, which together created the warp, it was possible to run a cross thread between them. The block of wood used to carry the filling strand through the warp was called the shuttle. The spinning wheel spun wool, while the loom wove cloth.

As we improved our machines, our own population was growing fast, matching the pace of change we set ourselves. There were between 2 and 6 million people in Roman Britain; between 1750 and 1851 the figure reached 21 million. By this time we were inventing machines, usually powered by humans or beasts, that could do complex tasks. Interestingly, the mechanized clock was one of our first complex machines. Improving machines and powering them were at the very heart of our next stage of development.

By the eighteenth century demand for clothing was such that the pressure was on to find a way of producing

material quickly. A breakthrough came in about 1764, when James Hargreaves is said to have come up with the idea for a hand-powered multiple spinning machine after observing a spinning wheel that his young daughter Jenny had accidentally overturned. When the spindle continued to revolve in an upright rather than a horizontal position, Hargreaves had the bright idea that many spindles could

A Newcomen colliery winding engine at Coalbrook, Shropshire, built around 1790. Engines based on Newcomen's principle were first constructed in 1712 as practical working steam engines for pumping water out of coal mines, and were widely used in industrial centres across Europe.

powering it was eventually found using a waterwheel. At the beginning of this period, the major sources of power available to industry were wind and water, either of which could turn the wheels and cogs that powered spindles, looms and grindstones. By the middle of the nineteenth century the refinement of waterwheel construction prepared the way for the water turbine, which is still an extremely efficient device for converting energy.

Steam became the characteristic and most commonly used power source of the British Industrial Revolution. The Newcomen atmospheric steam engines had been installed for pumping water out of coal mines, but little else. Only modest development took place in the original atmospheric engine until James Watt patented a separate condenser in 1769. It did not simply replace other sources of power – it transformed them. From that point onwards, the steam engine underwent almost continuous improvements for more than a hundred years.

Over the next quarter of a century, Watt and his partner Boulton converted the steam engine from a single-acting atmospheric pumping machine, which applied power only on the downward stroke of the piston, into a double-acting piston that could drive a wheel in a rotary manner, thus moving the wheels of industry.

A British textile manufacturer, Sir Richard Arkwright, adopted the rotary action engine for use in his cotton mill. The results were astonishing: 714 spindle mules could glide across the floor spinning 1.6 kilometres (1 mile) of yarn every 20 seconds. It also showed the potential of applying steam power to large-scale grain milling. Many other industries followed in exploring the possibilities of steam power, and it soon became widely used.

Machines that could be powered continuously required workers to operate them. Factories were built close to trading centres – which were, of course, cities. The modern city was finally born as people moved from

be so turned at once. He built a machine on which one person could spin several threads at a time. In the time it had previously taken to produce one spindle, the Spinning Jenny could now produce six.

The next step was to find a way to power this simple machine and so produce even more spindles of thread. After many variations on the basic design a way of the countryside, where fewer people were required because steam tractors and other machines were doing more of the work, to obtain employment in the new factories. Times were set for the start and end of work. Productivity was at last being set against time.

By the end of the eighteenth century, populations of cities were expanding fast. And this new mobility from country to town brought with it another change. Transport was undergoing its own revolution, speeding up travel and conquering distance.

How travel changed the world

Until the middle of the nineteenth century travelling from one place to another, even if the distance was no more than a few kilometres, was arduous and time-consuming. Unless you were rich and owned a horse, you walked. Most people rarely moved out of the village in which they were born. But with new cities growing, people had to move and a solution was desperately needed.

One of the most important developments of the Industrial Revolution was the railway. The invention of the steam engine meant that for the first time we could generate enough power to make something as large as a train move. The Cornish engineer Richard Trevithick introduced higher steam pressures in 1802 with an experimental engine at Coalbrookdale, which worked safely and efficiently. He then applied his engine to a vehicle, making the first successful steam locomotive for the Penydarren tram road in South Wales in 1804. This was, however, more of a technical triumph than a commercial one, because the locomotive fractured the cast-iron track of the tramway. The age of the railway had to await further improvements.

A very early locomotive, the Steam Elephant from 1810, had numerous cogwheels and rods, but eventually refinements made steam trains economical to run. Following the opening of the Stockton and Darlington Railway in 1825, the cities of Liverpool and Manchester decided to build a 64.5-kilometre (40-mile) railway connecting them. George Stephenson was delegated the task of constructing the line using his famous 'Rocket'. For a short stretch the locomotive achieved a speed of

58 kilometres (36 miles) per hour and convinced people that train travel was commercially viable. When Isambard Kingdom Brunel completed the railway line from London to Bristol the journey could be covered in just over four hours – before that a horse and carriage would have taken some 16 hours.

In America the railroad, as it was known, became an essential part of westward expansion for people seeking a new life. The first line was built in the 1830s from Baltimore to Ohio and thereafter the railroad quickly expanded everywhere. Without it, it is probably true to say that there would be no unified United States of America. The impact railways had on the distance people and freight could cover was enormous, and the importance of the time saved cannot be overestimated. Whereas it would have taken four months to travel from east coast to west by wagon trail, once the rail link was completed in 1869 it took just four days. California was suddenly not that far away.

Energy was the key to travelling more quickly and over longer distances. With each development in the railways, bigger and faster engines were required and that meant more fuel. This was costly, but the more people were prepared to use this method of transport, the cheaper it became. Travel was taking on a new meaning as it became more accessible to everyone.

But the railways in America did much more to time. In order to set proper timetables, time zones had to be established for each state. When the clocks went back in the autumn (fall), trains would stop for a whole hour so that they arrived at the next station at the time written in the timetable. In spring when they lost an hour the trains had to make up the time by going more quickly!

Across the world, trains dominated long-distance travel until the Second World War, when they began to lose out to aeroplanes. However, before that, another development had revolutionized the concept of independent travel.

The birth of the modern engine

The internal-combustion engine emerged in the nineteenth century as a result of both greater scientific understanding of the principles of thermodynamics and a search by engineers for a substitute for the large steam

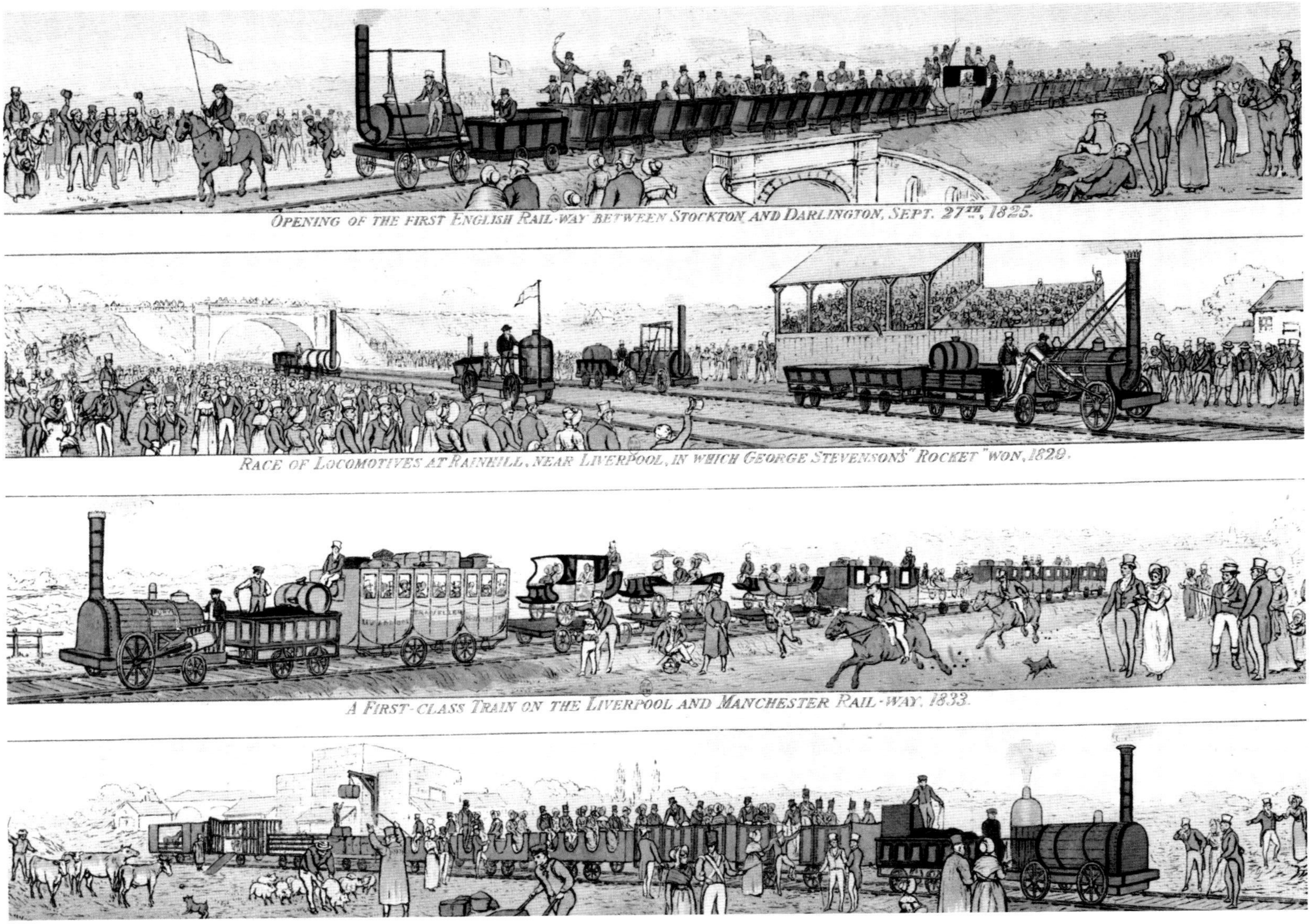

OPENING OF THE FIRST ENGLISH RAIL-WAY BETWEEN STOCKTON AND DARLINGTON, SEPT. 27TH, 1825.

RACE OF LOCOMOTIVES AT RAINHILL, NEAR LIVERPOOL, IN WHICH GEORGE STEVENSON'S "ROCKET" WON, 1829.

A FIRST-CLASS TRAIN ON THE LIVERPOOL AND MANCHESTER RAIL-WAY, 1833.

engine. In an internal-combustion engine the fuel is burned inside the engine. The original idea was based on the cannon, which provided an early model of a single-stroke engine. Several people had experimented with gunpowder as a means of driving a piston in a cylinder.

The first successful gas engine was made by Étienne Lenoir in Paris in 1859. It used an explosive mixture of gas and air ignited by an electric spark on alternate sides of the piston when it was in mid-stroke position. Although this worked quite well, it was costly to operate and it was not until a refinement introduced by the German inventor Nikolaus Otto in 1878 that the gas

An illustration showing the advent of train travel in the nineteenth century. From top to bottom: the opening of the Stockton and Darlington Railway in 1825; the Rainhill trials of 1820, won by Stephenson's Rocket; the Liverpool and Manchester first-class train in 1833; and the Liverpool and Manchester second-class train in 1833.

engine became a commercial success. Otto adopted the four-stroke cycle of induction-compression-firing-exhaust. As a result, gas engines came to be used extensively in small industrial establishments.

Further modification was necessary to make the engine more mobile and, with the development of refined

Henry Ford's conveyor-belt system of 1913 increased the speed of car production. Not only could this improved process keep up with the demand for new cars, it also made them more affordable.

petroleum, a substitute fuel was at last found. With the internal-combustion engine came the motor car, which gave the individual the possibility of independent travel. Until now, only horse and coach could do that, and keeping a horse was difficult and time-consuming. The motor car put an end to that. You could go whenever and wherever you wanted to go. Given the right roads and tyres, plus a good engine, a car could move far quicker than a horse or even ten horses. Over the years, although reliability and safety, comfort and efficiency have greatly improved, the basic design of a motor car has changed very little.

It was in the United States with its burgeoning industrial base and long distances that the popularity of the motor car first took off. Demand was so high that a way of mass-producing them had to be found. Henry Ford's conveyor-belt system, introduced in Detroit in 1913, increased production by rolling off a new car every 10 seconds. This was achieved by having each worker in one set place, doing only one job.

From continent to continent

Mass migration around the world during the nineteenth century was done on ships. Major breakthroughs in design meant that ships could set sail on time and arrive on time. Most importantly, they were quicker.

Brunel made outstanding contributions to marine engineering with his three important ships, the *Great Western* in 1837, *Great Britain* in 1843 and *Great Eastern* in 1858, each the largest in the world at its date of launching. The *Great Western*, a wooden paddle vessel, was the first steamship to provide a regular transatlantic service. The *Great Britain*, an iron-hull steamship, was the first large vessel driven by a screw propeller and also provided transatlantic passenger service. The *Great Eastern* was propelled by both paddles and screw and was the first ship to utilize a double iron hull. Unsurpassed in size for 40 years, the *Great Eastern* was not a success as a passenger ship but achieved fame by laying the first transatlantic cable, an important feat in the development of communication.

Taking to the air

After the Second World War both ships and trains were surpassed by air travel, which resulted in their decline, but not their obliteration. The birth of propeller-driven aircraft and especially jet aircraft reduced the time it took to get across oceans. While a journey by ship from Southampton to Wellington, New Zealand, still takes four weeks, a jet aircraft can do it in 26 hours. It is now estimated that the longest it would take anyone to travel to anywhere in the world is 48 hours. That is an astonishing thought, considering that just a hundred years ago we simply had not been everywhere on the planet, with both North and South Poles yet to be set foot on. Yet by 1969 we had set foot on a different planet – the moon.

So super-powered machines speeded up the time it takes us, and goods, to move about the Earth, and no doubt we will make transport even quicker in the future. In the process of conquering distance, we have also conquered the time it takes to cover it.

As we saw in the introduction to this book, Einstein calculated that the quicker we travel, the greater the distortion of time. The traveller's time is slower than that of the stationary person and so he returns slightly younger than he would have been if he had stayed put. The passing of time is the same, but the traveller has been away slightly longer from the point of view of the people he left behind. With faster travelling and greater distances conquered in space, this distortion will lead to greater discrepancies in time in the future.

By the beginning of the twentieth century, factories, cities and transport were having a major impact on the landscape and on our lifestyle. There was access to so much more, not only in places to go but in goods to buy, and the population was relentless in its growth. Demand for food, clothes and luxury increased. The age of consumerism had begun.

Mass production leads to mass graves for our unwanted material goods. These B52 bombers parked in the Arizona desert have outlived the purpose for which they were originally constructed, and are a clear indicator of how our consumerism and waste are changing the face of the Earth, perhaps forever.

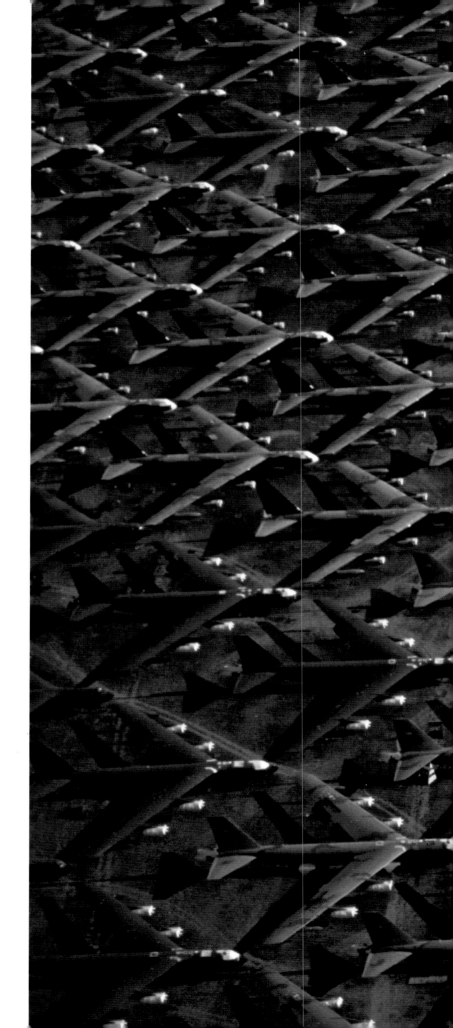

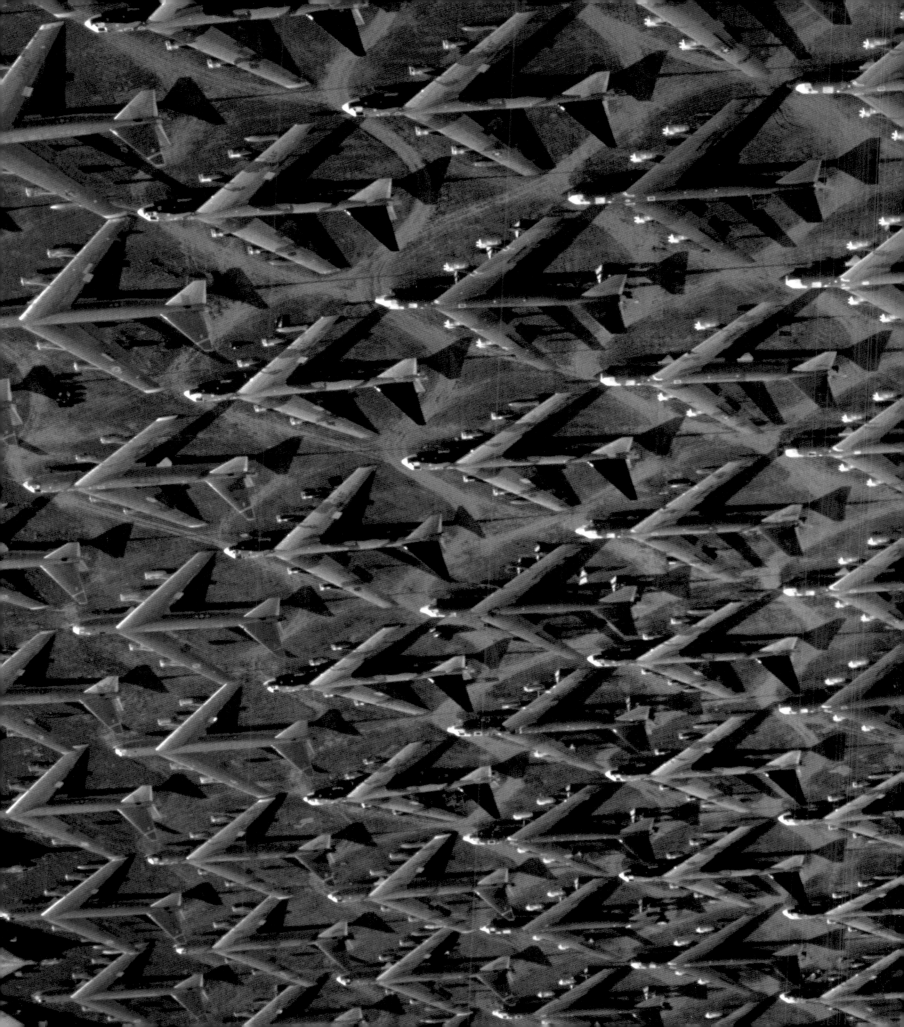

The year-round tomato

With the growth of industry came two new aspects to our lives: we had more money to spend, and more free time. We could spend our spare cash on goods we wanted. The motor car was one, but fashionable clothes, music albums and going to the movies were the real consumer's choice. The array and abundance of food in huge supermarkets was another. The demand for certain items is so great that we now want what used to be seasonal food, such as asparagus and strawberries, all year around. Even the tropical tomato has gone through a complete change.

The tomato has become one of the most popular fruits of the modern age, surpassed only by the apple and the banana. A large percentage of the world's crop is used for processing, to make such products as canned tomatoes, tomato juice, ketchup, purée, paste and 'sun-dried' tomatoes. Consumption in Britain has increased threefold over 25 years. With almost 60 million people, we are eating some three and a half billion more tomatoes than we were in 1980. Such is the demand that we have had to find new ways of producing them – fast.

The wild species originally came from the Andes of South America, probably Peru and Ecuador. It was later domesticated in Mexico as an ornamental plant long before the arrival of Europeans. The Spanish then introduced the tomato to Europe in the early sixteenth century and, with the Italians, were the first people to adopt it as a food. It has, of course, remained a staple of Mediterranean cuisine.

In northern Europe the tomato was also first grown as an ornamental. Botanists acknowledged it as a relative of the poisonous belladonna and deadly nightshade, so it was regarded with great suspicion as a foodstuff. The roots and leaves of the tomato plant are in fact poisonous, containing the neurotoxin solanine.

Those first tomatoes were not bright red but a golden yellow colour, which inspired one of the first European names, *pomi d'oro*, Italian for golden apple. Since then it has been estimated that over 10,000 different strains of tomatoes have been bred from the five wild varieties originally found in South America.

Growing the tomato in temperate northern climates required ways of intensifying both heat and light, as well as protecting the plants from frost. The nineteenth century saw the birth of the greenhouse, which not only changed how tomatoes were grown but also where they were grown. Later large-scale glasshouses were introduced that could accommodate thousands of tomato plants. Tomatoes could now be grown commercially anywhere in the world. Even Iceland can produce its own, using heat from volcanic vents.

Today in Britain about 300 hectares (740 acres) of glasshouses are used to produce tomatoes. That's about 530 football pitches, all full of tomatoes. At Runcton in

Sussex, one of Britain's biggest tomato-producing areas, there are 9 hectares (23 acres) of glasshouses dedicated to the tomato. This artificial environment is precision-controlled on every level. Heat, light, water, pest control, pollination and nutrition are all geared to rearing the ultimate tomato.

The glasshouse has become a sort of Time Machine, producing more tomatoes more quickly and more efficiently than is possible in the wild. Whatever the weather outside, the temperature inside is controlled at a steady 21°C (70°F), 24 hours a day, seven days a week. A boiler keeps things hot and vents cool them down. When there's too much sun, shade protects the leaves from

Many vegetables, like the tomatoes pictured here, are now grown in large-scale glasshouses that enable their temperature to be controlled at all times. Not only do they mature much faster in this environment, but they can also be made available to consumers throughout the year.

scorching. On average each person working in the industry is producing three times more tomatoes per hour than they were 25 years ago. Delivery is fast, too – it takes from one to three days from the time it is harvested for a British tomato to reach the supermarket shelf.

Water, as we know, is crucial to agriculture and the ways it is harnessed can be ingenious. The problem arises when we fail to predict the demand placed on such a valuable source.

Leaving ships high and dry

To visit the Aral Sea in Kazakhstan is to see one of the most amazing examples of what we can do to a huge environment and a precious reservoir for wildlife. If you do not know what to expect you could be in for a horrendous shock. Several kilometres before you get to the sea you will see huge ships lying on their side on dry land, abandoned and rusting away. The land you are standing on was once covered by the sea, but the shore is now hundreds of kilometres away. The sea has simply shrunk in alarming proportions.

In the early years of the twentieth century, the Aral Sea covered an area roughly the size of Ireland, making it the world's fourth largest inland body of water. It was an excellent place to catch fish. Then in the 1960s a vast waterway network was developed in the area for the cultivation of crops, such as cotton and rice. These upstream irrigation schemes have starved the Aral Sea of its water. It is no wonder – to satisfy the thirsty cotton, the fields were criss-crossed with 700,000 kilometres (437,500 miles) of channels using water that used to go to the Aral Sea. The cotton fields now cover roughly the same amount of land as the sea, about 70,000 square kilometres (27,500 square miles).

In a Time Machine you can see the speed at which these changes occur. First the water currents of the Amu Darya and Syr Darya rivers, which feed the Aral Sea, dwindle to a disturbing degree, and then the sea very quickly loses 50 per cent of its area and a massive 75 per cent of its water. As you watch, the sea level falls by an astonishing 14 metres (46 feet), about the size of a four-storey house.

As the water has diminished, its salinity has increased continually in the course of the past 30 years, reaching about three times its original concentration and causing the disappearance of more than 20 species of fish. On top of that, the increased salt carried by the winds burns all the vegetation within a radius of hundreds of kilometres, contributing to the desertification of the environment.

The phenomenon of the Aral Sea, although among the best known, is not unique. Some 400,000 square kilometres (156,000 square miles) of irrigated land all over the world have excessive salt, largely due to man's irrigation schemes.

Changing our landscape

The speed of change in the countryside today is giving us a very distorted view of the world. Whereas the environmental changes were too slow for the Aborigines in Australia to realize what was going on, today they are sometimes too quick for us to react to in time. We are simply not geared up to call a halt quickly enough, especially if things are going disastrously wrong, as they did with the Aral Sea.

Modern farming has kept pace with the changes of cities and technology. The mechanization of the twentieth century meant that big machines needed big fields, which quickly led to the demise of hedgerows in many parts of Britain. Across the world the emphasis on mass production has produced the same effect in the landscape – monoculture. That is because a single crop grown on a large scale is simply quicker to harvest than anything more diverse.

The pressure on the countryside has not only come from farming. The construction of cities and machines demanded ever more resources. Iron was mined in vast quantities but soon many other metals, such as tin, copper and aluminium, were required, leading us to exploit the riches of our planet still further. And this required us to think of ways in which we could acquire enough of each resource quickly and on time. We had to think big in order to power our new, consumption-driven fast-paced lifestyle.

Vanishing sea

Intense irrigation of the land surrounding the Aral Sea in Kazakhstan has resulted in a dramatic change to the landscape. Within just a few decades, the sea has shrunk to a mere shadow of its former size – it used to be the world's fourth largest inland body of water. The process was so quick that many ships were left high and dry.

12 years ago

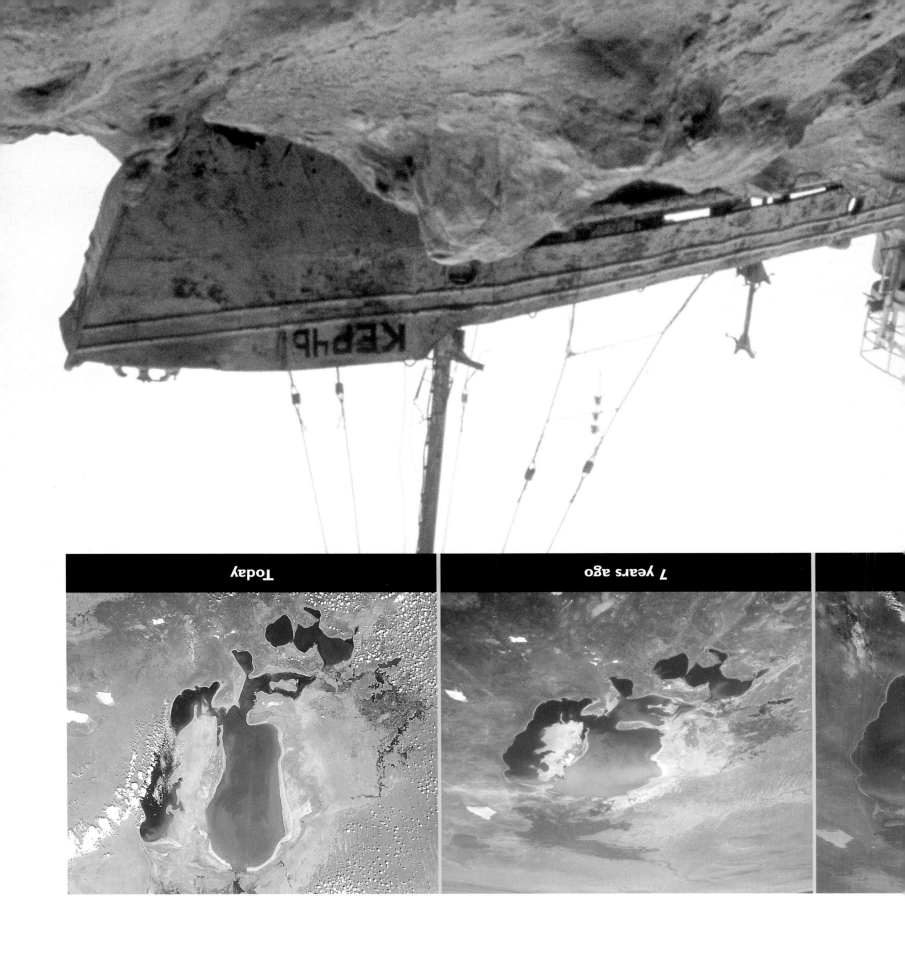

Massive mines

How quickly the landscape around Hamersley Iron in Western Australia has changed can be seen over just 30 years.

30 years ago

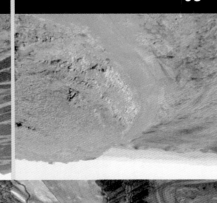

Tapping the Earth's resources

Two types of resources were in demand: energy in the form of coal, oil and gas; and the materials for building, both rock and metals. Industrial and domestic consumption demanded that exploitation be done on a large scale. Starting with vast coal mines, many deep within the Earth, powerful machines helped us extract valuable commodities.

When we started digging into the Earth we used hand tools. It was hard work, but it was effective. If we wanted to dig a bigger hole to find minerals, we simply employed more people to dig. If we wanted to speed up the process, we did the same.

But in the past 150 years, manpower has been taken over by machine power, and in the process we have moved mountains – literally. Extraordinarily, the total amount of earth and rock we have shifted has actually been measured. It is estimated to be equivalent to a mountain range 4000 metres (over 13,000 feet) high, 100 kilometres (62 miles) long and 40 kilometres (25 miles) wide, and most of that has happened since the Industrial Revolution in the mid-eighteenth century. Hamersley Iron in Western Australia is just one of many huge mines around the world being dug to provide

Hamersley Iron in Western Australia is the largest iron ore mine in the world. Once a landscape of undulating plains, the speed at which the ore has been extracted since its original discovery has altered the face of this area beyond recognition. This is one change that we can clearly see happening within a human lifetime.

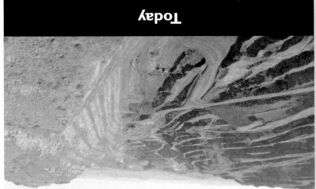

20 years ago   10 years ago   Today

us with the raw materials to build cities, roads, cars, bicycles, motorbikes, toys, jewellery – in fact most of the things we manufacture. It is 6 kilometres (4 miles) long, 300 metres (1000 feet) deep and it gets bigger every day. It operates 24 hours a day, seven days a week, to provide the iron ore that is used to make the steel that builds our planes, trains and the gigantic trucks that carry the ore away from the mine in the first place. Each truckload is over 200 tonnes of ore, which is taken to the crusher and broken down into smaller bits. In a single day these machines move more earth than a single man and his spade could move if he worked every hour of every day for a whole year. Put another way, one man and his machine have speeded up the process of changing the face of the planet.

Nearly 70 million tonnes of iron ore are dug out every year from this one part of Western Australia. The rocks from this mine are so rich in metal that they can be welded directly onto other metal. This high quality means the iron is in demand around the world. Destined for places as far away as the UK, it begins its journey in trains up to 2 kilometres (1¼ miles) long with three diesel engines hauling 200 carriages, each of which carries 100 tonnes of ore. That's enough iron to construct about 15,000 family cars.

Nature might be a very powerful force, but when it comes to eating away at the Earth's surface, we humans can be far more destructive. We are the most active landscapers the world has ever seen. Today 37 billion tonnes of earth are shifted every year, so we really do have the power to make the Earth move, both on a large scale and with speed.

Gone are the days of spending days and weeks hunting and gathering food from the wild. Today we want it all in an instant. Technology helps achieve some things a lot more quickly, but it has also provided us with something we now take totally for granted: instant power and communication.

The power of electricity

Electricity as a source of power was developed in the middle of the nineteenth century. The pioneering work

had been done by an international collection of scientists including Benjamin Franklin of Pennsylvania, Alessandro Volta of the University of Pavia, Italy, and Michael Faraday of Britain. It was Faraday who demonstrated the nature of the relationship between electricity and magnetism in 1831, and his experiments provided the point of departure for both the mechanical generation of electric current, previously available only from batteries, and the employment of such a current in electric motors. Both the mechanical generator and the motor depend on the rotation of a continuous coil of conducting wire between the poles of a strong magnet: turning the coil produces a current, while passing a current through the coil causes it to turn. Both generators and motors underwent substantial development in the middle decades of the nineteenth century. In particular, French, German, Belgian and Swiss engineers evolved the most satisfactory forms of the wire coil and produced the dynamo, which made the large-scale generation of electricity commercially possible.

The next problem was to find a market for this new resource. In the United States, Thomas Edison applied his inventive genius to finding fresh uses for electricity, and his development of the carbon-filament lamp showed how this form of energy could rival gas lamps as a domestic source of light. Edison and the English chemist Sir Joseph Swan experimented independently with various materials for the filament and both chose carbon. The result was a highly successful small lamp, which could be varied in size and power depending on the purpose for which it was intended.

Lighting alone could not provide an economical market for electricity because its use was confined to the hours of darkness. Successful commercial generation depended on the development of other uses, and particularly on electric traction. The popularity of urban electric tramways and the adoption of electric traction

The lights of Times Square are one of the most famous attractions of New York City, but large quantities of electrical power are essential for its continuing success. When the city experienced a total black-out on 14 August 2003, the local residents said it felt just like a return to medieval times.

on subway systems, such as the London Underground, coincided with the widespread construction of generating equipment in the late 1880s and 1890s. The subsequent spread of this form of energy is one of the technological success stories of the twentieth century, although most of the basic techniques of its generation, distribution and utilization had been mastered by the end of the nineteenth century.

During the 1800s water and gas became available on demand in people's homes. All we had to do was turn a tap, but the uses to which these luxuries could be put were very limited and only the rich had access to them. But with the introduction of electricity came a rush of inventions that, within the space of a mere hundred years, changed the way we lived our lives for ever.

Electric power has become one of the greatest consumers of our age. For everyday living we use it from the time we get up in the morning to the time we go to bed. From washing machines to vacuum cleaners, instruments powered by electricity help us with our household chores. Electricity also powers our personal time- and labour-saving devices, from shavers to toothbrushes. All these inventions offer us an easier lifestyle, with more time to spare.

Instant power, instant communication

The development of electrical power and communication was to distort our perception of time and distance all over the world. Distance had always been the great barrier. Trains, planes and motorcars were now moving us fast from one location to another, but there was another hurdle imposed by distance and that could be overcome only by improved communication. In the mid-eighteenth century a letter was the only way to send information and it took as long as the person carrying it and his horse took to cover the journey. By the middle of the nineteenth century that was all to change.

In addition to improved transport and domestic convenience, there was another fundamental development of electrical power: telegraphy and the telephone, using cable technology. Starting with dots and dashes in the form of Morse code, named after the man

who invented it, it was not long before our own voices could travel down the telegraph lines. Distance was no longer a factor. We could talk to someone thousands of kilometres away as if they were right beside us.

Then came public radio and later television as a means of broadcasting voice and pictures without wires. This technology was an instant window to a bigger world. Communication was seen as power and in the early days it was controlled by the state. But radio and television were soon to be outstripped by another, even bigger technology.

Instant and constant communication, like the mobile cellphone, has transformed our lives forever. We can now link up to anywhere in the world at the touch of a button, which makes our vast planet seem so much smaller and more manageable and saves us so much time as well.

The personal computer has taken over our lives. Most of us now have one in our homes, enabling us to send written messages, view pictures and communicate with people on the other side of the planet, all at the click of a button. The Internet, which has grown so rapidly only in the last ten years, is the new form of communication. It is a largely unregulated source of information on which you can see, hear and read anything you wish – if you can find it. Instant communication and information-sourcing has finally gripped our world like a net.

The next breakthrough was digital technology, which has now infiltrated every electrical device from the mobile phone to television and radio. We are no longer tied to a land line: we use satellite phones, even videophones, from anywhere on the globe – all in an instant. But, as we know from experience, the computer and electronics world relies on power; without it, it ceases to function.

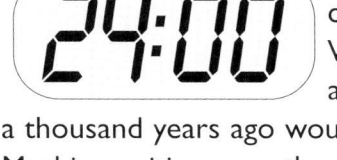 In one day in the twenty-first century, a human living in the West consumes the same amount of energy as a person a thousand years ago would have consumed in a year. Machines, cities, even the countryside are run on electricity. From space we can see how the Earth is lit by the energy we create. Our landscapes are littered with machines that convert raw resources into the energy we need, from nodding oil donkeys to hydroelectric dams. Digging, drilling and burning fuel, plus the huge industrial buildings that are associated with it, create some of the worst eyesores known, but more importantly they can all have a long-term environmental impact.

Alternative energy

Could there be an alternative to these blots on our landscape, and if so what would the landscapes of the future be like? Where could the Time Machine take us and what would we see? The quest is on to harness natural energy, such as the power of the wind and the sun. Could the windmill return as a serious energy supplier? We can conceive a landscape transformed by a sea of thousands of windmills, all making electricity from a clean, natural source – moving air. But this is not a vision of the future – it already exists. Today, there are over 13,000 of these giant towers of power spread throughout the UK and it's the fastest growing form of renewable energy.

Wind as a source of power is a real alternative to poisonous fossil fuels; wind 'farms' of thousands of

windmills are already supplying electricity straight to the US national grid. In the future they could become a common sight around the world. And instead of covering our deserts with concrete jungle, perhaps we could cover them with a sea of mirrors. Enough energy comes from the sun every day to power our homes and businesses for years to come – the problem is tapping into this potential. But in the sun-drenched Mojave Desert a plant developed in the mid-1980s now provides electricity to the many people of California.

Mirrors can be used to concentrate the heat from the sun to create solar-thermal electricity. To make sure they get the most of the sun's energy they are programmed to follow its movement across the sky. Mirrored parabolic troughs focus the sun's energy to an oil-filled receiver tube, which uses it to superheat the oil to a temperature of over 300°C (570°F). The heat from the oil then produces electricity using the more traditional nineteenth-century method of a steam-turbine generator. It turns water into steam, which drives the generators and produces electricity. The whole production process is monitored by computers to make sure all goes smoothly. It is no small operation. The shimmering solar farm covers over 400 hectares (1000 acres) and has 500,000 mirrors covering 1,000,000 square metres (90,000 square feet). Every day these massive reflectors are checked for breakages and reflectivity so that none of the sun's energy is wasted. Solar power here in the Californian desert is a reality, but it remains to be seen whether it will catch on elsewhere.

Wind and solar farms may be harnessing natural sources of energy, but to many people they are just as much of an eyesore as any open-cast mine or power station. Whatever the resources of the future, the way we use them will always have an impact on the landscape in some way or another. Whatever we do, wherever we go,

Massive solar panels in the Mojave Desert, USA, harness the sun's power to satisfy energy needs. There are real concerns that, in just a few decades, fossil fuels will be exhausted due to our increasing demand, so sights like this may become more familiar in the future as we search for new sources of sustainable energy.

A moutain of disposable bottles in the Philippines. Although these bottles have been collected for recycling, similar piles of waste are one of the unfortunate products of our consumer society. Much packaging is only useful for a very short space of time, but once abandoned can take hundreds of years to decompose. We must find ways to deal with such waste as human populations grow.

we'll always make an impression. The problem is the time we are taking to make those impressions.

What used to take humans generations to change and create, we can now do in a lifetime or even in just a few years. Things that used to take months can be done in an instant. With the right equipment we can contact any place in the world. Messages and information travel at the speed of light, the fastest anything can travel. Communication and its instant 24/7 coverage of events around the world have seen the expansion of television and radio channels and the Internet. There is access to so much data that there is little time to analyze exactly what is going on. Our perception of the world has been distorted by a constant flood of information.

We not only take on local problems in our immediate world, we now take on everything on a global level. Now the only things with which we can easily identify are global icons, like pop stars and sports personalities, and global brands that sell on a massive scale.

Has crushing time to the point where everything is happening in an instant made our world more superficial? Or are we capable, now more than ever, of obtaining information that will enable us to create a better world? Maybe we are creating our own 'world', seeking only those we want to seek, talking only to people we can relate to and rejecting the others. Or will we do the opposite – become more tolerant, listen more, take everyone's dreams and problems on board and try to solve them for the sake of humanity and the planet we live in?

Throughout our history, we humans have tried to understand the space and time in which we live. We define the boundaries of our lives and invent ways of overcoming them. Our overactive inventive minds play with ideas, developing theories that can become fact through technological invention. Astronomers are charting the stars, cartographers the Earth, our physical boundaries are being drawn, all with greater accuracy. New forensic technology means we can look back in time to the beginning of creation; it also allows us to look into the future to find solutions to present-day problems. Still our view of our existence has not been resolved. We have not got all the answers.

Technology not only measures time more precisely, it saves us time. Human populations and consumption of resources are increasing, thanks to our ability to utilize time. If we continue to live as we currently are, we will need two planets, not just the one we have.

Into the future

So what kind of future do we have? What could our Time Machine reveal if we took a trip into the future? Over the next few decades it is possible that we will buy more time for ourselves by increasing our potential average lifespan. Healthy controlled diets, better lifestyles and new medicines will help us achieve this. But there is a downside. As we live longer the burden on our resources will increase. Human populations are due to double every decade and cities and transportation will grow in proportion to this, especially in developing countries, such as Africa and Asia. Wild places, currently untouched by humans, will quickly disappear. Food and fuel production will have to speed up to sustain the growing population. Sadly it does not look like our planet will be able to maintain these high demands. So how can we reverse the trend towards the destruction of our planet?

Understanding the Earth and its nature is fundamental to understanding our own future. Nature's time is so different to our time. Yet we are finding ways of speeding up events originally set by the cosmic cycle of the solar system. We have tapped natural resources so that we can do things instantly. We have started to play with time, by altering the rate of change, and we can manipulate progress for our own ends. Technology has got us here, but in the end it could be new technologies with a better understanding of our planet that will save us.

Over time I believe that we will discover more secrets to life at a molecular level and it is there that we will find our solutions to population growth, food supply and even saving species and habitats. New energy will be harnessed safely at an atomic level to power our industries. But will our ever-improving technology meet the new challenges thrown up by our runaway progress, or will we be too slow to respond; indeed, are we too late?

Only time will tell.

In making both this extraordinary TV series and a book about time, there has been a host of people who helped transformed the original idea of the Time Machine into a reality. From those who worked on the early concept within the Development Office in the BBC TV Natural History Unit (NHU) to those who spent nearly two years on this project, as well as those who joined the team on either work experience or short-term contracts, I am extremely thankful. This idea did not come from just one person but flourished as a result of the many participants who saw its potential. There are numerous people, too many to mention, who gave wonderful comments and suggestions along the way, and I am extremely grateful to all of them.

I would like to express my appreciation of Keith Scholey, who was then Head of the NHU and offered me this challenging television series. I am also thankful to Sara Ford, Executive Producer, and Neil Nightingale, Head of the NHU, for their helpful suggestions during the making of the series, some of which are also reflected in the book.

I would like to give special thanks to those who helped me directly while I wrote this book. In particular, I express my sincere gratitude to the producers Scott Alexander, Anuschka de Rohan, Adam White and Dan Rees, for their help in hammering out the stories based on the enormous amount of research carried out. The majority of the research for the series, and on which the book depended, was the result of the hard work of our researchers, particularly Laura Fudge, Ann Stubbs, Caroline Marriott and Ingrid Kivale.

I am extremely grateful to the Television Production Coordinators, Jolie Bradfield and Gaynor Scattergood, who made such a great effort in every way they could to make sure the whole production ran smoothly. I am also thankful to Sally Mark, Production Manager, who kept an eye on the overall production finances, including Graphics.

There were many scientists who helped us with their great expertise and knowledge. I am particularly indebted to Simon Lamb and Chris Searle from Oxford University for help with the Earth chapter, and Ian Wilson at the University of Auckland and Brad Scott of Geological Nuclear Sciences in New Zealand, who pointed me in the right direction for earthquakes and volcanoes. I am very grateful for the help Farouk el-Baz from Boston University gave on the Sphinx, Stephen Stokes from Oxford University on the Sahara, and Andrew Warren and Andrew Goudie at Oxford University for information on wind erosion. For the Life chapter, I would like to thank Rae Silver from Columbia University for sharing his knowledge on circadian clocks, John Cooley for his expertise on cicadas, Pete Davie from Queensland Museum on soldier crabs, Tim Hill for his work on yew trees and Ethan Temeles on St Lucia's hummingbird story. As regards the Human chapter, I am very grateful to John Macgee and Gifford Miller from the University of Newcastle for the Australian fire story. On the perception of time I would like to thank John Wearden at the University of Manchester and on body clocks Russell Foster at Imperial College London.

For discovering the stunning pictures in this book, I wish to thank Frances Abraham, and for the book's wonderful design, Martin Hendry. I would especially like to thank Jan Golunski and his creative team at Jelly Television, Bristol, who produced superb graphics for the series as well as the book, particularly the innovative 'deep time lapse' sequences of changing landscapes and evolving animals.

I would also like to thank Shirley Patton, Editorial Director at BBC Books, who gave me the chance to write my first book and I would like to express my sincere gratitude to Sarah Miles for her editorial help and advice, and for guiding me through the whole process with great ease. I am also very thankful to Caroline Taggart and Ben Morgan for their assistance and suggestions that improved my writing and the sense of the manuscript enormously.

I would like to thank my wife, Maggi, and daughter, Sasha, for their support on this project and many others in the past, especially when I was away for long periods. I would also like to thank my friends who encouraged me along the way.

Finally I am indebted to both my late parents, who encouraged me in my interest in the natural sciences, which eventually led to my career making programmes in the best television department in the world – the BBC's Natural History Unit.

Bernard Walton

BBC Worldwide would like to thank the following for providing photographs and for permission to reproduce copyright material. While every effort has been made to trace and acknowledge all copyright holders, we would like to apologize should there have been any errors or omissions.

Page 1 FLPA (NASA); 2–3 Corbis (Alan Schein Photography); 4 Bernard Walton/BBC; 6 Science Photo Library (D. Weintraub); 8–9 Science Photo Library (Eadweard Muybridge Collection, Kingston Museum); 12 Corbis (Tom & Dee Ann McCarthy); 13 Nature Picture Library (Anup Shah); 14 Ancient Art & Architecture Collection; 15 Salisbury Cathedral (Steve Day); 17 Science & Society; 18 NHPA (Laurie Campbell); 20–1 Science Photo Library (Larry Landolfi); 22–3 Getty Images (Hulton Archive); 24–5 Ronald Grant Archive; 26–7 Bernard Walton/BBC; 28 Katz (George Steinmetz); 29 Bernard Walton/BBC; 30–1 OSF (Mark Hamblin); 32 Nature Picture Library (Simon King); 35 Science Photo Library (François Gohier); 36–7 Science Photo Library (Royal Observatory, Edinburgh, AAO); 38–9 all Bernard Walton/BBC; 40–1 Corbis (Michael S. Yamashita); 42 Bernard Walton/BBC; 43 Getty Images (AFP); 44–5 all Meg Smith, Dept. of Geology, University of Auckland; 47 Nature Picture Library (Staffan Widstrand); 51 main picture Science Photo Library (NASA); 52 Science Photo Library (David Parker); 54–5, 56 bottom, 59, 60 & 61 Bernard Walton/BBC; 64 & 65 Corbis (Pierre Colombel); 67 OSF (NASA); 68 Science Photo Library (NASA); 69 Dan Rees/BBC; 70 NHPA (Karl Switak); 71 Science Photo Library (John Walsh); 72–3 all OSF; 74 OSF (Mark Hamblin); 75 Science Photo Library (Plailly/Euralios); 76 NHPA (Dave Watts); 78–9 Ardea (Jean-Paul Ferrero); 80–1 all bottom Nature Picture Library (Jose B. Ruiz); 81 top & 82–3 all Nature Picture Library (Stephen Dalton); 84 Ardea (P. Morris); 85 OSF (Owen Newman); 87 all FLPA (Silvestris); 88 OSF (Kathie Atkinson); 89 Nature Picture Library (Michael Pitts); 90 left OSF (Mike Slater); 90 right Natural Visions; 91 FLPA (Frans Lanting/Minden Pictures); 92 Peter J. Herring; 94 Corbis (Joe McDonald); 97 FLPA (Alan Parker); 98 top Auscape (Jean-Paul Ferrero);

100–1 FLPA (Konrad Wothe/Minden Pictures); 102–3 top Nature Picture Library (G. & H. Denzau); 104–5 Nature Picture Library (Anup Shah); 106 OSF (Michael Fogden); 108–9 Nature Picture Library (Brandon Cole); 111 main picture FLPA (David Fleetham/Silvestris); 111 inset Natural History Museum, London; 113 NHPA (Daniell Heuclin); 114 Scott Alexander/BBC; 115 Nature Picture Library (Jeremy Walker); 116–7 left Bernard Walton/BBC; 117 right Corbis (Paul Edmondson); 118 Corbis (Michael Macor/San Francisco Chronicle); 119 OSF (Clive Bromhall); 120–1 Science Photo Library (Pascal Goetgheluck); 123 NHPA (Christophe Ratier); 124 OSF (Gallo Images); 126 both left & 126–7 both insets Gaynor Scattergood/BBC; 127 main picture Scott Alexander/BBC; 130–1 OSF (David Cayless); 132 Scott Alexander/BBC; 135 Still Pictures/UNEP (Voltchev); 136–7 Corbis (Yann Arthus-Bertrand); 138 main picture Anuschka de Rohan/BBC; 141 Getty Images (Image Bank); 142–3 Science & Society; 145 Bridgeman Art Library (Transport Museum, London); 146–7 Getty Images (Hulton Archive); 148–9 Corbis (Joseph Sohm/ChromoSohm Inc.); 150–1 Holt Studios; 152–3 top left & 153 centre top NASA; 153 top right NHPA; 152–3 bottom Science Photo Library (Novosti Press Agency); 154–5 Auscape (Jean-Paul Ferrero); 156–7 Corbis (Alan Schein Photography); 158–9 Corbis (Randy Faris); 160–1 Scott Alexander/BBC; 162 Still Pictures/UNEP (J. Tanodra).

Unless credited otherwise above, the pictures of time-lapse sequences are © BBC.